T0297446

Research and Perspectives in Endocrine Interactions

More information about this series at http://www.springer.com/series/5241

Paolo Sassone-Corsi • Yves Christen
Editors

A Time for Metabolism and Hormones

 Springer

OPEN

Editors
Paolo Sassone-Corsi
Department of Biological Chemistry
University of California
Irvine, California
USA

Yves Christen
Fondation IPSEN
Boulogne-Billancourt Cedex, France

ISSN 1861-2253   ISSN 1863-0685   (electronic)
Research and Perspectives in Endocrine Interactions
ISBN 978-3-319-27068-5   ISBN 978-3-319-27069-2   (eBook)
DOI 10.1007/978-3-319-27069-2

Library of Congress Control Number: 2015957954

Springer Cham Heidelberg New York Dordrecht London

Printed on acid-free paper

Springer International Publishing AG Switzerland is part of Springer Science+Business Media (www.springer.com)

Preface

Each morning we wake up from a night of sleep, and each day we eat our regularly timed meals, go through our normal routines, and fall asleep again for another night. This rhythm, so-called circadian—after the Latin words circa diem ("about a day")—underlies a wide variety of human physiological functions, including sleep–wake cycles, body temperature, hormone secretion, exercise activity, and feeding behavior. Circadian rhythms are remarkably conserved throughout evolution, and it is becoming commonly appreciated that circadian rhythms represent an exquisite example of systems biology.

At the heart of all cyclic biological functions is the circadian clock, a highly conserved molecular system that enables organisms to adapt to common daily changes, such as the day–night cycle and food availability. The mammalian anatomical structure in the brain that governs circadian rhythms consists of a small area of the anterior hypothalamus, called the suprachiasmatic nucleus (SCN). For decades, this "central pacemaker" was thought to be the unique circadian clock of the organism. This dogma was challenged when peripheral tissues were also found to contain functional circadian oscillators that are self-sustained at the single cell level. This notion, together with the discovery that a remarkable fraction of the genome is transcriptionally controlled by the clock, illustrated that circadian control must play a key role in governing the metabolism and physiology of all organisms. This concept was recently validated by studies of the metabolome revealing that a large fraction of metabolites oscillate in a given tissue.

Recent years have seen spectacular advances in the field of circadian biology. These have attracted the interest of researchers in many fields, including endocrinology, neurosciences, cancer, and behavior. By integrating a circadian view within the fields of endocrinology and metabolism, researchers will be able to reveal many, yet-unsuspected aspects of how organisms cope with changes in the environment and subsequent control of homeostasis.

The concept behind the Fondation IPSEN *Colloque Medecine et Recherche* on "A Time for Metabolism and Hormones," held in Paris on December 5, 2014, was to capture the excitement of this field as it is opening new avenues in our understanding of metabolism and endocrinology. A panel of the most distinguished investigators in the field gathered together to discuss the present state and the future of the field. These proceedings constitute a compendium of the most updated views by these investigators. We trust that it will be of use to those colleagues who will be picking up the challenge to unravel how the circadian clock can be targeted for the future development of specific pharmacological strategies toward a number of pathologies.

Irvine, CA, USA   Paolo Sassone-Corsi
Boulogne-Billancourt Cedex, France   Yves Christen

Acknowledgements

The editors wish to express their gratitude to Mrs. Mary Lynn Gage for her editorial assistance and Mrs. Astrid de Gérard for the organization of the meeting.

Acknowledgements

Contents

List of Contributors

Lorena Aguilar-Arnal Department of Biological Chemistry, Center for Epigenetics and Metabolism, Unit 904 of INSERM, University of California, Irvine, CA, USA

Joseph Bass Department of Medicine, Division of Endocrinology, Metabolism and Molecular Medicine, Feinberg School of Medicine, Chicago, IL, USA

Leonardo Bee Department of Biological Chemistry, Center for Epigenetics and Metabolism, Unit 904 of INSERM, University of California, Irvine, CA, USA

Marion Benabou Institut Pasteur, Human Genetics and Cognitive Functions Unit, Paris, France

CNRS UMR3571 Genes, Synapses and Cognition, Institut Pasteur, Paris, France

Sorbonne Paris Cité, Human Genetics and Cognitive Functions, University Paris Diderot, Paris, France

Thomas Bourgeron Human Genetics and Cognitive Functions Unit, Institut Pasteur, Paris, France

CNRS UMR3571 Genes, Synapses and Cognition, Institut Pasteur, Paris, France

Sorbonne Paris Cité, Human Genetics and Cognitive Functions, University Paris Diderot, Paris, France

FondaMental Foundation, Créteil, France

Gillberg Neuropsychiatry Centre, Sahlgrenska Academy, University of Gothenburg, Gothenburg, Sweden

Steven A. Brown Chronobiology and Sleep Research Group, Institute of Pharmacology and Toxicology, University of Zürich, Zürich, Switzerland

Marlene Cervantes Department of Biological Chemistry, Center for Epigenetics and Metabolism, Unit 904 of INSERM, University of California, Irvine, CA, USA

Abhishek Chatterjee Institut de Neurosciences Paris-Saclay, CNRS/Université Paris Sud, Gif-sur-Yvette, France

Ludmila Gaspar Chronobiology and Sleep Research Group, Institute of Pharmacology and Toxicology, University of Zürich, Zürich, Switzerland

Jonathan Gaucher Center for Epigenetics and Metabolism, Unit 904 of INSERM, University of California, Irvine, CA, USA

Carla B. Green Department of Neuroscience, University of Texas Southwestern Medical Center, Dallas, TX, USA

Guillaume Huguet Human Genetics and Cognitive Functions Unit, Institut Pasteur, Paris, France

CNRS UMR3571 Genes, Synapses and Cognition, Institut Pasteur, Paris, France

Sorbonne Paris Cité, Human Genetics and Cognitive Functions, University Paris Diderot, Paris, France

Kenichiro Kinouchi Department of Biological Chemistry, Center for Epigenetics and Metabolism, Unit 904 of INSERM, University of California, Irvine, CA, USA

Mitchell A. Lazar Division of Endocrinology, Diabetes, and Metabolism, Department of Medicine, Department of Genetics, and The Institute for Diabetes, Obesity, and Metabolism, Perelman School of Medicine at the University of Pennsylvania, Philadelphia, PA, USA

Stafford Lightman Henry Wellcome Laboratories for Integrative Neuroscience and Endocrinology, University of Bristol, Bristol, UK

Selma Masri Department of Biological Chemistry, Center for Epigenetics and Metabolism, Unit 904 of INSERM, University of California, Irvine, CA, USA

Emilie Montellier Department of Biological Chemistry, Center for Epigenetics and Metabolism, Unit 904 of INSERM, University of California, Irvine, CA, USA

Mari Murakami Department of Biological Chemistry, Center for Epigenetics and Metabolism, Unit 904 of INSERM, University of California, Irvine, CA, USA

Ricardo Orozco-Solis Department of Biological Chemistry, Center for Epigenetics and Metabolism, Unit 904 of INSERM, University of California, Irvine, CA, USA

Akhilesh B. Reddy Department of Clinical Neurosciences, University of Cambridge Metabolic Research Laboratories, Cambridge, UK

NIHR Biomedical Research Centre, Wellcome-MRC Institute of Metabolic Science, University of Cambridge, Cambridge, UK

François Rouyer Institut de Neurosciences Paris-Saclay, CNRS/Université Paris Sud, Gif-sur-Yvette, France

Paolo Sassone-Corsi Department of Biological Chemistry, Center for Epigenetics and Metabolism, Unit 904 of INSERM, University of California, Irvine, CA, USA

Amita Sehgal Perelman School of Medicine, University of Pennsylvania, Philadelphia, PA, USA

Joseph S. Takahashi Howard Hughes Medical Institute, University of Texas Southwestern Medical Center, Dallas, TX, USA

Department of Neuroscience, University of Texas Southwestern Medical Center, Dallas, TX, USA

Paola Tognini Department of Biological Chemistry, Center for Epigenetics and Metabolism, Unit 904 of INSERM, University of California, Irvine, CA, USA

The Epigenetic and Metabolic Language of the Circadian Clock

Paolo Sassone-Corsi

Abstract The circadian clock controls a large variety of neuronal, endocrine, behavioral and physiological responses in mammals. This control is exerted in large part at the transcriptional level on genes expressed in a cyclic manner. A highly specialized transcriptional machinery based on clock regulatory factors organized in feedback autoregulatory loops governs a significant portion of the genome. These oscillations in gene expression are paralleled by critical events of chromatin remodeling that appear to provide plasticity to circadian regulation. Specifically, the NAD^+-dependent deacetylases SIRT1 and SIRT6 have been linked to circadian control of gene expression. This and additional accumulating evidence shows that the circadian epigenome appears to share intimate links with cellular metabolic processes and has remarkable plasticity, showing reprogramming in response to nutritional challenges. In addition to SIRT1 and SIRT6, a number of chromatin remodelers have been implicated in clock control, including the histone H3K4 tri-methyltransferase MLL1. Deciphering the molecular mechanisms that link metabolism, epigenetic control and circadian responses will provide valuable insights towards innovative strategies of therapeutic intervention.

Introduction

Metabolism, homeostatic balance and behavior follow the 24-h daily cycle (Eckel-Mahan and Sassone-Corsi 2013). Circadian rhythms are virtually present in all life forms on our planet, including mammals, insects, plants, fungi and cyanobacteria. In higher organisms, circadian rhythms have evolved into a complex physiological and molecular system demonstrated by sleep-wake cycles, daily fluctuations in body temperature, blood pressure, cellular regeneration and behavior such as food

P. Sassone-Corsi (✉)
Department of Biological Chemistry, Center for Epigenetics and Metabolism, Unit 904 of INSERM, University of California, Irvine, Irvine, CA 92697, USA
e-mail: psc@uci.edu

© The Author(s) 2016
P. Sassone-Corsi, Y. Christen (eds.), *A Time for Metabolism and Hormones*,
Research and Perspectives in Endocrine Interactions,
DOI 10.1007/978-3-319-27069-2_1

1

intake and alertness levels (Asher and Sassone-Corsi 2015). Metabolism, nutritional intake and body homeostasis are also under circadian control, displaying rhythms in the levels of circulating hormones and metabolites, as well as enzymes within the biochemical pathways participating in their biosynthesis (Eckel-Mahan and Sassone-Corsi 2013; Gamble et al. 2014). Circadian rhythms are so intimately linked to biological processes that their misregulation may lead to a number of pathologies such as obesity, metabolic syndrome, diabetes, cardiovascular diseases, inflammation, sleep disorders and some cancers (Eckel-Mahan and Sassone-Corsi 2013).

The molecular bases of circadian rhythms have been explored, revealing a remarkable variety of molecular mechanisms that underlie clock function. An important system of circadian control utilizes the core clock molecular machinery that consists of transcription factors and regulators, both activators and repressors, that act in concert to drive circadian expression of an important fraction of the genome. A number of high-throughput transcriptome profiling studies have established that 15–30 % of all transcripts are controlled by the clock, depending on the tissue or cell type (Duffield et al. 2002; Panda et al. 2002; Storch et al. 2002; Ueda et al. 2002). Accumulating evidence has shown that this global program of gene expression is achieved through events of cyclic chromatin remodeling and epigenetic control.

Chromatin Remodeling, Cyclic Transcription and the Clock

The molecular organization of the circadian system relies on a network of cellular oscillators present in virtually every cell of the organism. An intricate network of transcriptional-translational feedback loops constitutes the molecular clock (Eckel-Mahan and Sassone-Corsi 2013; Zhang and Kay 2010). The basic helix-loop-helix (b-HLH)-PAS proteins CLOCK and BMAL1 are core elements of this system and function as transcriptional activators to drive the expression of many clock-controlled genes (CCGs). CLOCK and BMAL1 heterodimers bind E-boxes in CCG promoters and activate their expression. Among the CCGs there are genes encoding other core clock protein repressors Period (PER1-3) and Cryptochromes (CRY1-2). PER and CRY proteins heterodimerize in the cytoplasm and translocate to the nucleus to inhibit CLOCK:BMAL1-mediated transcription. The stability of PER:CRY complexes is regulated by posttranscriptional modifications (Lee et al. 2009) and ubiquitination events (Busino et al. 2007; Hirano et al. 2013; Siepka et al. 2007; Yoo et al. 2013). The time-controlled clearance of the repressors primes for a the next cycle of CLOCK:BMAL1-driven gene activation. This system then leads to the cyclic activation of other regulatory pathways generating interconnected transcriptional feedback loops. These provide remarkable plasticity to the circadian system, eliciting multiple daily oscillations in the transcriptome (Masri and Sassone-Corsi 2010).

Specific cyclic chromatin transitions occur in a genome-wide scale and are associated with circadian waves of transcription (Masri and Sassone-Corsi 2010). Several chromatin remodelers have been found to be involved in circadian control. The protein CLOCK was found to operate as an acetyltransferase on histone H3 at K9 and K14 (Doi et al. 2006), modifications associated with a chromatin state permissive for transcription. CLOCK acts in concert with other histone acetyltransferases (HATs) (Etchegaray et al. 2003), such as CBP (CREB binding protein), p300 and with the CBP-associated factor PCAF (Lee et al. 2010; Curtis et al. 2004; Takahata et al. 2000). A number of histone deacetylases (HDACs) have been found to counterbalance these HATs. For example, the circadian repressor PER recruits SIN3A-HDAC1 (Duong et al. 2011), whereas the protein CRY1 associates with the complex SIN3B-HDAC1/2 (Naruse et al. 2004). The circadian regulator REV-ERBα recruits the NCoR-HDAC3 complex in a rhythmic manner to chromatin via a process that has been linked to the control of lipids metabolism in the liver (Sun et al. 2011). Thus, a variety of circadian repressive complexes appear to exist that may elicit distinct functions at unique times of the circadian cycle. The nicotinamide adenine dinucleotide (NAD^+)-dependent class III of HDACs was found to play a critical role in connecting cellular metabolism to circadian physiology. The founding member, SIRT1, gives the name to this class of enzymes, collectively known as sirtuins. There are seven sirtuins, all involved in various aspects of metabolism, inflammation and aging; their intracellular localization is nuclear, cytoplasmic or mitochondrial. The nuclear proteins SIRT1 and SIRT6 have been shown to contribute to circadian transcription (Nakahata et al. 2008; Masri et al. 2014).

A number of chromatin post-translational modifications have been linked to clock function in addition to acetylation. The first evidence that a histone modification may play a role in circadian transcription was the light-inducible phosphorylation at H3-S10 in SCN neurons (Crosio et al. 2000). The activating histone methylation H3K4me3 has also been linked to clock control and it seems to be essential to permit circadian chromatin transitions that lead to activation of CCG expression (Ripperger and Schibler 2006). MLL1, a H3K4 histone methyltransferase (HMT), was shown to elicit CLOCK:BMAL1 recruitment to chromatin at specific circadian promoters and for the cyclic tri-methylation at H3K4 (Katada and Sassone-Corsi 2010). Also the repressive mark H3K27me3 is clock controlled at the *Per1* promoter through a mechanism that involves the methyltransferase EZH2 (Etchegaray et al. 2006). Additional chromatin remodelers involved in circadian function include the demethylase JARID1a that appears to inhibit HDAC1, thereby enhancing CLOCK:BMAL1-mediated transcription (DiTacchio et al. 2011), and the FAD (Flavin Adenine Dinucleotide)-dependent demethylase LSD1 whose function is controlled by PKCα-mediated circadian phosphorylation (Nam et al. 2014).

Cellular Metabolism and the Circadian Clock Converge

A large number of human studies and animal models provide solid evidence of the reciprocal regulation between the circadian clock and cellular and organismal homeostasis (Eckel-Mahan and Sassone-Corsi 2013; Dallmann et al. 2012; Eckel-Mahan et al. 2012, 2013; Hatori et al. 2012; Kasukawa et al. 2012). The clock regulates metabolism by controlling the expression of a large fraction of the genome. Moreover, the oscillator appears to sense the cellular energy state and consequently adapts its function accordingly.

Several levels of interplay exist between cellular metabolism and chromatin remodeling (Masri and Sassone-Corsi 2010; Feng and Lazar 2012; Katada et al. 2012). Acetylation of histones or non-histone nuclear proteins depends on the supply of acetyl-CoA in the nuclear compartment. The main carbon source in mammals is glucose, which generates acetyl-CoA because of the enzyme adenosine triphosphate (ATP)-citrate lyase (ACLY). ACLY protein levels are cyclic in the liver (Mauvoisin et al. 2014), and ACLY activity controls global histone acetylation depending on glucose availability (Wellen et al. 2009). Thus, circadian changes in histone acetylation are controlled not only by specific HATs but also by interconnected metabolic pathways and enzymes supplying nuclear acetyl-CoA. A similar regulation involves S-adenosyl methionine (SAM), the metabolite used by methyltransferases to deliver methyl groups. Changing SAM levels directly influence H3K4me3 levels in mouse pluripotent stem cells (Shyh-Chang et al. 2013). Also, treatment with 3-deazaadenosine (DAA), an inhibitor of SAH (S-adenosylhomocysteine) hydrolysis that hinders transmethylation, elongates the circadian period (Fustin et al. 2013). Further research is necessary to decipher the impact of one carbon metabolism in the circadian transcriptome.

Nicotinamide adenine dinucleotide (NAD^+) is a pivotal metabolite for the circadian epigenome. NAD^+ shows robust diurnal rhythms in synchronized cells and mice (Bellet et al. 2013; Nakahata et al. 2009; Ramsey et al. 2009), and operates as a cofactor for class III of HDACs, the sirtuins (see next section).

The core machinery may be directly influenced by changing metabolic states. Specifically, the DNA-binding function of NPAS2:BMAL1 and CLOCK:BMAL1 heterodimers was shown to be influenced by the redox states of NAD(H) or NADP (H) (Rutter et al. 2001). This finding implied that CLOCK:BMAL1 transcriptional activity should be sensitive to the levels of cellular redox. While a causal evidence for this regulation has not been explored, circadian oscillations in intracellular redox potentials are evolutionary conserved (Eckel-Mahan and Sassone-Corsi 2013; Asher and Sassone-Corsi 2015). Thus, while the ability of NPAS2 or CLOCK to sense the intracellular redox state in vivo remains to be proven, independent evidence provides interesting information. Indeed, crystallographic analyses of the CRY1-PER2 complex indicate that a disulfide bond between two cysteine residues in CRY1 weakens its interaction with PER2, whereas a reduced state of CRY1 stabilizes the complex and facilitates transcriptional repression (Schmalen et al. 2014). In this scenario, CRY2 would retain specific FAD (Flavin

Adenine Dinucleotide) binding activity, and FAD competes for CRY2 binding pocket with the ubiquitin ligase complex SCF^{FBXL3}, which has been shown to control period length by regulating CRYs stability (Xing et al. 2013). Interestingly, this finding provides a possible approach to pharmacologically adjust circadian period length by using small molecules resembling FAD (Hirota et al. 2012).

Posttranslational modifications of clock proteins have been shown to modify their regulatory capacity. For example, CLOCK, BMAL1 and PER2 can be O-linked N-acetylglucosamine (GlcNAc)-modified by the enzyme O-GlcNAc transferase (OGT), which results in a change in their activities (Kaasik et al. 2013; Li et al. 2013). Importantly, liver-specific ablation of OGT leads to dampened oscillation of *Bmal1* and gluconeogenic genes. Thus, glucose levels dictate the availability of GlcNAc, OGT serving as a signal transducer between cellular metabolism and circadian components. Along the same lines, phosphorylation of CRY1 by the nutrient sensor kinase AMPK (AMP-activated protein kinase) connects cellular energy levels with the circadian clock by adjusting it to the changing intracellular ratio of AMP/ATP (Jordan and Lamia 2013; Gomes et al. 2013).

The Central Role of Sirtuins

The intracellular availability in time and space of specific metabolites constitutes an intriguing level of control for their protein sensors (Katada et al. 2012). In this respect, the circadian oscillation in NAD^+ concentration represents a revealing paradigm. The NAD^+ biosynthetic salvage pathway controls the conversion of nicotinamide (NAM) to β-nicotinamide mononucleotide (NMN); this step is catalyzed by a rate-limiting step enzyme, the nicotinamide phosphoribosyltransferase (NAMPT, also known as visfatin). The circadian machinery controls the transcription of the *Nampt* gene through direct binding of CLOCK:BMAL1 to E-boxes in the promoter (Nakahata et al. 2009; Ramsey et al. 2009). NMN is converted to NAD^+ by the enzymes nicotinamide mononucleotide adenylyltransferase 1-3 (NMNAT1-3) (Fig. 1). Thus, a transcriptional-enzymatic feedback loop controls NAD^+ biosynthesis and availability that in turn could result in circadian function of a variety of NAD^+-dependent enzymes. Moreover, there is a differential regulation of NAD^+ levels and NAD^+-consuming enzymes in various cell compartments (Gomes et al. 2013; Yang et al. 2007). In this respect the sirtuins deserve special attention. Indeed, of the seven mammalian sirtuins, three (SIRT1, SIRT3 and SIRT6) have been functionally linked to circadian control and found to modulate cyclic outputs in response to metabolic cues.

SIRT3 is a mitochondrial enzyme that displays robust changes in its deacetylase activity in response to NAD^+ levels (Hebert et al. 2013; Peek et al. 2013; Masri et al. 2013). SIRT3 controls mitochondrial function, including fatty acid oxidation and intermediary metabolism, by directly targeting rate-limiting enzymes for mitochondrial biochemical processes (Peek et al. 2013). As mitochondrial fatty acid

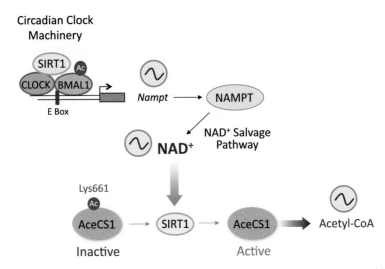

Fig. 1 Metabolism and the circadian clock converge. A paradigm example is represented by the role of SIRT1 and other sirtuins in clock regulation. The circadian machinery controls a large fraction of the genome through the transcriptional regulation of CCGs. One of the CCGs is the gene encoding the protein NAMPT, the rate-limiting enzyme in the NAD^+-salvage pathway. Cyclic transcriptional control of the *Nampt* gene results in the cyclic synthesis of NAD^+, which in turn is consumed rhythmically by enzymes such as SIRT1, whose deacetylase activity is consequently cyclic. One of the non-histone targets is the enzyme AceCS1, which contributes to the synthesis of Acetyl-CoA. AceCS1 is acetylated at one residue, Lys661, and its cyclic deacetylation by SIRT1 activates the enzyme, resulting in cyclic synthesis of Acetyl-CoA and thereby oscillating availability of acetyl groups required for global acetylation

oxidation and protein acetylation show circadian rhythmicity (Masri et al. 2013), the link with NAD^+ availability through SIRT3 is of particular interest. Also, mitochondria from *Bmal1$^{-/-}$* mice display reduced oxidative ability and decreased mitochondrial NAD^+ levels (Peek et al. 2013). These findings, together with the implication of SIRT1 in circadian control, raise the possibility that the sirtuins-NAD^+ link with the clock may represent a critical molecular pathway to govern the process of aging.

The implication of nuclear sirtuins in clock function is multiple. SIRT1 is both nuclear and cytoplasmic whereas SIRT6 is exclusively nuclear and mostly chromatin bound, localized at transcriptionally active genomic loci. SIRT1 and SIRT6 operate through distinct mechanisms to coordinate the clock machinery in a differential manner and thereby delineate the circadian transcriptional output (Masri et al. 2014). Because of these different mechanisms of action, in the liver these two sirtuins coordinate circadian expression of distinct groups of genes. SIRT6 exerts its function by coordinating CLOCK:BMAL1 recruitment to specific chromatin sites (Masri et al. 2014). SIRT1, which is mostly nucleoplasmic and is recruited to chromatin only 'on demand', deacetylates histones and non-histone proteins. Among the non-histone targets of SIRT1 there are the clock proteins BMAL1 and PER2 (Asher et al. 2008; Hirayama et al. 2007). SIRT1 is also able to deacetylate MLL1, thereby controlling its methyltransferase activity. Thus, there is control in H3K4 tri-methylation through the cyclic oscillation of NAD$^+$ levels (Aguilar-Arnal et al. 2015).

SIRT1-mediated deacetylation also affects circadian levels of other metabolites besides NAD$^+$. Specifically, intracellular acetyl-CoA levels are controlled by the clock through SIRT1-controlled deacetylation of the enzyme acetyl-CoA Synthetase 1 (AceCS1) (Sahar et al. 2014). This acetylation switch controls AceCS1 activity, leading to cyclic synthesis of acetyl-CoA (Fig. 1), that then is likely to influence the acetylation levels of histones and non-histone proteins (Sahar et al. 2014). In contrast, SIRT6 deacetylase activity seems to be efficient in removing long chain fatty acids from lysine residues (Jiang et al. 2013). In this respect it is noteworthy that not only on NAD$^+$, but also on fatty acids, control the activity of SIRT6 (Feldman et al. 2013). Thus, SIRT6 appears to occupy a key position in the control of fatty acids metabolism by the clock. Indeed, CLOCK: BMAL1-driven activation of genes involved in fatty acid biosynthesis is modulated by SIRT6 (Masri et al. 2014).

High-throughput analysis of the transcriptome and metabolome along the circadian cycle has revealed notable differences in the metabolic functions of SIRT1 and SIRT6. Using mice with liver-specific deletion of either SIRT1 or SIRT6, a specific role for SIRT6 was shown in dictating the synthesis and breakdown of fatty acid pathways, as well as their storage into triglycerides. SIRT6 operates at least in part through the control of alternative circadian transcriptional pathways, specifically because of the chromatin recruitment of the sterol regulatory element-binding protein 1 (SREBP1) (Masri et al. 2014). Thus, it is through genomic partitioning that the two deacetylases contribute to a parallel segregation of cellular metabolism (Masri et al. 2014).

Finally, these findings suggest a role for genome topology in circadian control (Aguilar-Arnal et al. 2013). Our studies have identified the presence of circadian interactomes where co-regulated genes are physically associated in the circadian epigenome. Nuclear sirtuins may constitute a paradigm for other chromatin remodelers that could contribute in the cyclic control of the nuclear landscape. Also, specific changes in the nuclear localization of NAD$^+$ may provide the possibility of restricting the distribution of this metabolite to "niches" of activity (Katada et al. 2012).

Conclusion

The ability of the circadian clock machinery to sense the metabolic state of the cell in a time-specific manner places it in a strategic position. Indeed, fascinating findings reviewed in this article demonstrate the direct implication of the clock in the maintenance of cellular homeostasis. The clock machinery appears to integrate environmental and metabolic signals to directly translate them in plasticity in gene expression so to favor the adaptation of the organism to specific conditions. As the circadian transcriptional landscape is highly complex, including dynamic changes in nuclear organization (Katada et al. 2012; Aguilar-Arnal et al. 2013), it becomes critical to decipher how the nuclear landscape integrates metabolic cues and shapes the transcriptional output. It is through the analysis of the specific coordination that key chromatin remodelers have with clock transcription factors that we will gain insights into how the intracellular metabolic state communicates with the clock machinery. As disruption of clock function has been linked to a variety of pathological conditions, revealing the clock mechanisms will lead to innovative strategies towards the pharmacological treatment of metabolic syndromes, obesity, diabetes, inflammation and even cancer.

Acknowledgments Studies in the Center for Epigenetics and Metabolism are supported by the National Institute of Health, Merieux Research Grants, and INSERM (Institut National de la Sante et Recherche Medicale, France).

References

Aguilar-Arnal L, Hakim O, Patel VR, Baldi P, Hager GL, Sassone-Corsi P (2013) Cycles in spatial and temporal chromosomal organization driven by the circadian clock. Nat Struct Mol Biol 20:1206–1213

Aguilar-Arnal L, Katada S, Orozco-Solis R, Sassone-Corsi P (2015) NAD$^+$-SIRT1 control of H3K4 trimethylation through circadian deacetylation of MLL1. Nat Struct Mol Biol 22:312–318

Asher G, Sassone-Corsi P (2015) Time for food: the intimate interplay between nutrition, metabolism and the circadian clock. Cell 161:84–92

Asher G, Gatfield D, Stratmann M, Reinke H, Dibner C, Kreppel F, Mostoslavsky R, Alt FW, Schibler U (2008) SIRT1 regulates circadian clock gene expression through PER2 deacetylation. Cell 134:317–328

Bellet MM, Nakahata Y, Boudjelal M, Watts E, Mossakowska DE, Edwards KA, Cervantes M, Astarita G, Loh C, Ellis JL, Vlasuk GP, Sassone-Corsi P (2013) Pharmacological modulation of circadian rhythms by synthetic activators of the deacetylase SIRT1. Proc Natl Acad Sci USA 110:3333–3338

Busino L, Bassermann F, Maiolica A, Lee C, Nolan PM, Godinho SI (2007) SCFFbxl3 controls the oscillation of the circadian clock by directing the degradation of cryptochrome proteins. Science 316:900–904

Crosio C, Cermakian N, Allis CD, Sassone-Corsi P (2000) Light induces chromatin modification in cells of the mammalian circadian clock. Nat Neurosci 3:1241–1247

Curtis AM, Seo SB, Westgate EJ, Rudic RD, Smyth EM, Chakravarti D, FitzGerald GA, McNamara P (2004) Histone acetyltransferase-dependent chromatin remodeling and the vascular clock. J Biol Chem 279:7091–7097

Dallmann R, Viola AU, Tarokh L, Cajochen C, Brown SA (2012) The human circadian metabolome. Proc Natl Acad Sci USA 109:2625–2629

DiTacchio L, Le HD, Vollmers C, Hatori M, Witcher M, Secombe J, Panda S (2011) Histone lysine demethylase JARID1a activates CLOCK-BMAL1 and influences the circadian clock. Science 333:1881–1885

Doi M, Hirayama J, Sassone-Corsi P (2006) Circadian regulator CLOCK is a histone acetyltransferase. Cell 125:497–508

Duffield GE, Best JD, Meurers BH, Bittner A, Loros JJ, Dunlap JC (2002) Circadian programs of transcriptional activation, signaling, and protein turnover revealed by microarray analysis of mammalian cells. Curr Biol 12:551–557

Duong HA, Robles MS, Knutti D, Weitz CJ (2011) A molecular mechanism for circadian clock negative feedback. Science 332:1436–1439

Eckel-Mahan K, Sassone-Corsi P (2013) Metabolism and the circadian clock converge. Physiol Rev 93:107–135

Eckel-Mahan KL, Patel VR, Mohney RP, Vignola KS, Baldi P, Sassone-Corsi P (2012) Coordination of the transcriptome and metabolome by the circadian clock. Proc Natl Acad Sci USA 109:5541–5546

Eckel-Mahan KL, Patel VR, de Mateo S, Orozco-Solis R, Ceglia NJ, Sahar S, Dilag-Penilla SA, Dyar KA, Baldi P, Sassone-Corsi P (2013) Reprogramming of the circadian clock by nutritional challenge. Cell 155:1464–1478

Etchegaray JP, Lee C, Wade PA, Reppert SM (2003) Rhythmic histone acetylation underlies transcription in the mammalian circadian clock. Nature 421:177–182

Etchegaray JP, Yang X, DeBruyne JP, Peters AH, Weaver DR, Jenuwein T, Reppert SM (2006) The polycomb group protein EZH2 is required for mammalian circadian clock function. J Biol Chem 281:21209–21215

Feldman JL, Baeza J, Denu JM (2013) Activation of the protein deacetylase SIRT6 by long-chain fatty acids and widespread deacylation by mammalian sirtuins. J Biol Chem 288:31350–31356

Feng D, Lazar MA (2012) Clocks, metabolism, and the epigenome. Mol Cell 47:158–167

Fustin JM, Doi M, Yamaguchi Y, Hida H, Nishimura S, Yoshida M, Isagawa T, Morioka MS, Kakeya H, Manabe I, Okamura H (2013) RNA-methylation dependent RNA processing controls the speed of the circadian clock. Cell 155:793–806

Gamble KL, Berry R, Frank SJ, Young ME (2014) Circadian clock control of endocrine factors. Nat Rev Endocrinol 10:466–475

Gomes AP, Price NL, Ling AJ, Moslehi JJ, Montgomery MK, Rajman L, White JP, Teodoro JS, Wrann CD, Hubbard BP, Mercken EM, Palmeira CM, de Cabo R, Rolo AP, Turner N, Bell EL, Sinclair DA (2013) Declining NAD(+) induces a pseudohypoxic state disrupting nuclear-mitochondrial communication during aging. Cell 155:1624–1638

Hatori M, Vollmers C, Zarrinpar A, DiTacchio L, Bushong EA, Gill S, Leblanc M, Chaix A, Joens M, Fitzpatrick JA, Ellisman MH, Panda S (2012) Time-restricted feeding without reducing caloric intake prevents metabolic diseases in mice fed a high-fat diet. Cell Metab 15:848–860

Hebert AS, Dittenhafer-Reed KE, Yu W, Bailey DJ, Selen ES, Boersma MD, Carson JJ, Tonelli M, Balloon AJ, Higbee AJ, Westphall MS, Pagliarini DJ, Prolla TA, Assadi-Porter F, Roy S, Denu JM, Coon JJ (2013) Calorie restriction and SIRT3 trigger global reprogramming of the mitochondrial protein acetylome. Mol Cell 49:186–199

Hirano A, Yumimoto K, Tsunematsu R, Matsumoto M, Oyama M, Kozuka-Hata H, Nakagawa T, Lanjakornsiripan D, Nakayama KI, Fukada Y (2013) FBXL21 regulates oscillation of the circadian clock through ubiquitination and stabilization of cryptochromes. Cell 152:1106–1118

Hirayama J, Sahar S, Grimaldi B, Tamaru T, Takamatsu K, Nakahata Y, Sassone-Corsi P (2007) CLOCK-mediated acetylation of BMAL1 controls circadian function. Nature 450:1086–1090

Hirota T, Lee JW, St John PC, Sawa M, Iwaisako K, Noguchi T, Pongsawakul PY, Sonntag T, Welsh DK, Brenner DA, Doyle FJ 3rd, Schultz PG, Kay SA (2012) Identification of small molecule activators of cryptochrome. Science 337:1094–1097

Jiang H, Khan S, Wang Y, Charron G, He B, Sebastian C, Du J, Kim R, Ge E, Mostoslavsky R, Hang HC, Hao Q, Lin HS (2013) SIRT6 regulates TNF-alpha secretion through hydrolysis of long-chain fatty acyl lysine. Nature 496:110–113

Jordan SD, Lamia KA (2013) AMPK at the crossroads of circadian clocks and metabolism. Mol Cell Endocrinol 366:163–169

Kaasik K, Kivimäe S, Allen JJ, Chalkley RJ, Huang Y, Baer K, Kissel H, Burlingame AL, Shokat KM, Ptáček LJ, Fu YH (2013) Glucose sensor O-GlcNAcylation coordinates with phosphorylation to regulate circadian clock. Cell Metab 17:291–302

Kasukawa T, Sugimoto M, Hida A, Minami Y, Mori M, Honma S, Honma K, Mishima K, Soga T, Ueda HR (2012) Human blood metabolite timetable indicates internal body time. Proc Natl Acad Sci USA 109:15036–15041

Katada S, Sassone-Corsi P (2010) The histone methyltransferase MLL1 permits the oscillation of circadian gene expression. Nat Struct Mol Biol 17:1414–1421

Katada S, Imhof A, Sassone-Corsi P (2012) Connecting threads: epigenetics and metabolism. Cell 148:24–28

Lee H, Chen R, Lee Y, Yoo S, Lee C (2009) Essential roles of CKIdelta and CKIepsilon in the mammalian circadian clock. Proc Natl Acad Sci USA 106:21359–21364

Lee Y, Lee J, Kwon I, Nakajima Y, Ohmiya Y, Son GH, Lee KH, Kim K (2010) Coactivation of the CLOCK-BMAL1 complex by CBP mediates resetting of the circadian clock. J Cell Sci 123:3547–3557

Li MD, Ruan HB, Hughes ME, Lee JS, Singh JP, Jones SP, Nitabach MN, Yang X (2013) O-GlcNAc signaling entrains the circadian clock by inhibiting BMAL1/CLOCK ubiquitination. Cell Metab 17:303–310

Masri S, Sassone-Corsi P (2010) Plasticity and specificity of the circadian epigenome. Nat Neurosci 13:1324–1329

Masri S, Patel VR, Eckel-Mahan KL, Peleg S, Forne I, Ladurner AG, Baldi P, Imhof A, Sassone-Corsi P (2013) Circadian acetylome reveals regulation of mitochondrial metabolic pathways. Proc Natl Acad Sci USA 110:3339–3344

Masri S, Rigor P, Cervantes M, Ceglia N, Sebastian C, Xiao C, Roqueta-Rivera M, Deng C, Osborne TF, Mostoslavsky R, Baldi P, Sassone-Corsi P (2014) Partitioning circadian transcription by SIRT6 leads to segregated control of cellular metabolism. Cell 158:659–672

Mauvoisin D, Wang J, Jouffe C, Martin E, Atger F, Waridel P, Quadroni M, Gachon F, Naef F (2014) Circadian clock-dependent and -independent rhythmic proteomes implement distinct diurnal functions in mouse liver. Proc Natl Acad Sci USA 111:167–172

Nakahata Y, Kaluzova M, Grimaldi B, Sahar S, Hirayama J, Chen D, Guarente LP, Sassone-Corsi P (2008) The NAD$^+$-dependent deacetylase SIRT1 modulates CLOCK-mediated chromatin remodeling and circadian control. Cell 134:329–340

Nakahata Y, Sahar S, Astarita G, Kaluzova M, Sassone-Corsi P (2009) Circadian control of the NAD$^+$ salvage pathway by CLOCK-SIRT1. Science 324:654–657

Nam HJ, Boo K, Kim D, Han DH, Choe HK, Kim CR, Sun W, Kim H, Kim K, Lee H, Metzger E, Schuele R, Yoo SH, Takahashi JS, Cho S, Son GH, Baek SH (2014) Phosphorylation of LSD1 by PKCalpha is crucial for circadian rhythmicity and phase resetting. Mol Cell 53:791–805

Naruse Y, Oh-hashi K, Iijima N, Naruse M, Yoshioka H, Tanaka M (2004) Circadian and light-induced transcription of clock gene Per1 depends on histone acetylation and deacetylation. Mol Cell Biol 24:6278–6287

Panda S, Antoch MP, Miller BH, Su AI, Schook AB, Straume M, Schultz PG, Kay SA, Takahashi JS, Hogenesch JB (2002) Coordinated transcription of key pathways in the mouse by the circadian clock. Cell 109:307–320

Peek CB, Affinati AH, Ramsey KM, Kuo HY, Yu W, Sena LA, Ilkayeva O, Marcheva B, Kobayashi Y, Omura C, Levine DC, Bacsik DJ, Gius D, Newgard CB, Goetzman E, Chandel NS, Denu JM, Mrksich M, Bass J (2013) Circadian clock NAD$^+$ cycle drives mitochondrial oxidative metabolism in mice. Science 342:1243417

Ramsey KM, Yoshino J, Brace CS, Abrassart D, Kobayashi Y, Marcheva B, Hong HK, Chong JL, Buhr ED, Lee C, Takahashi JS, Imai S, Bass J (2009) Circadian clock feedback cycle through NAMPT-mediated NAD$^+$ biosynthesis. Science 324:651–654

Ripperger JA, Schibler U (2006) Rhythmic CLOCK-BMAL1 binding to multiple E-box motifs drives circadian Dbp transcription and chromatin transitions. Nat Genet 38:369–374

Rutter J, Reick M, Wu LC, McKnight SL (2001) Regulation of clock and NPAS2 DNA binding by the redox state of NAD cofactors. Science 293:510–514

Sahar S, Masubuchi S, Eckel-Mahan K, Vollmer S, Galla L, Ceglia N, Masri S, Barth TK, Grimaldi B, Oluyemi O, Astarita G, Hallows WC, Piomelli D, Imhof A, Baldi P, Denu JM, Sassone-Corsi P (2014) Circadian control of fatty acid elongation by SIRT1 protein-mediated deacetylation of acetyl-coenzyme A synthetase 1. J Biol Chem 289:6091–6097

Schmalen I, Reischl S, Wallach T, Klemz R, Grudziecki A, Prabu JR, Benda C, Kramer A, Wolf E (2014) Interaction of circadian clock proteins CRY1 and PER2 is modulated by zinc binding and disulfide bond formation. Cell 157:1203–1215

Shyh-Chang N, Locasale JW, Lyssiotis CA, Zheng Y, Teo RY, Ratanasirintrawoot S, Zhang J, Onder T, Unternaehrer JJ, Zhu H, Asara JM, Daley GQ, Cantley LC (2013) Influence of threonine metabolism on S-adenosylmethionine and histone methylation. Science 339:222–226

Siepka SM, Yoo SH, Park J, Song W, Kumar V, Hu Y, Lee C, Takahashi JS (2007) Circadian mutant overtime reveals F-box protein FBXL3 regulation of cryptochrome and period gene expression. Cell 129:1011–1023

Storch KF, Lipan O, Leykin I, Viswanathan N, Davis FC, Wong WH, Weitz CJ (2002) Extensive and divergent circadian gene expression in liver and heart. Nature 417:78–83

Sun Z, Feng D, Everett LJ, Bugge A, Lazar MA (2011) Circadian epigenomic remodeling and hepatic lipogenesis: lessons from HDAC3. Cold Spring Harb Symp Quant Biol 76:49–55

Takahata S et al (2000) Transactivation mechanisms of mouse clock transcription factors, mClock and mArnt3. Genes Cells 5:739–747

Ueda HR, Chen W, Adachi A, Wakamatsu H, Hayashi S, Takasugi T, Nagano M, Nakahama K, Suzuki Y, Sugano S, Iino M, Shigeyoshi Y, Hashimoto S (2002) A transcription factor response element for gene expression during circadian night. Nature 418:534–539

Wellen KE, Hatzivassiliou G, Sachdeva UM, Bui TV, Cross JR, Thompson CB (2009) ATP-citrate lyase links cellular metabolism to histone acetylation. Science 324:1076–1080

Xing W, Busino L, Hinds TR, Marionni ST, Saifee NH, Bush MF, Pagano M, Zheng N (2013) SCF (FBXL3) ubiquitin ligase targets cryptochromes at their cofactor pocket. Nature 496:64–68

Yang H, Yang T, Baur JA, Perez E, Matsui T, Carmona JJ, Lamming DW, Souza-Pinto NC, Bohr VA, Rosenzweig A, de Cabo R, Sauve AA, Sinclair DA (2007) Nutrient-sensitive mitochondrial NAD+ levels dictate cell survival. Cell 130:1095–1107

Yoo SH, Mohawk JA, Siepka SM, Shan Y, Huh SK, Hong HK, Kornblum I, Kumar V, Koike N, Xu M, Nussbaum J, Liu X, Chen Z, Chen ZJ, Green CB, Takahashi JS (2013) Competing E3 ubiquitin ligases govern circadian periodicity by degradation of CRY in nucleus and cytoplasm. Cell 152:1091–1105

Zhang EE, Kay SA (2010) Clocks not winding down: unravelling circadian networks. Nat Rev Mol Cell Biol 11:764–776

Molecular Architecture of the Circadian Clock in Mammals

Joseph S. Takahashi

Abstract The circadian clock mechanism in animals involves an autoregulatory transcriptional feedback loop in which CLOCK and BMAL1 activate the transcription of the *Period* and *Cryptochrome* genes. The PERIOD and CRYPTOCHROME proteins then feed back and repress their own transcription by interaction with CLOCK and BMAL1. We have studied the biochemistry of the CLOCK:BMAL1 transcriptional activator complex using structural biology as well as the genomic targets of CLOCK and BMAL1 using ChIP-seq methods. We describe the dynamics of the core circadian clock transcriptional system. CLOCK and BMAL1 interact with the regulatory regions of thousands of genes. The gene network and dynamics of the system will be discussed. A mechanistic description of the core circadian clock mechanism should promote our understanding of how the circadian clock system influences behavior, physiology and behavioral disorders.

Introduction

Over the last 20 years, my laboratory has been focused on understanding the molecular mechanism of circadian clocks in mammals. We have used mouse genetics as a tool for discovery of the critical genes involved in the generation of circadian rhythms of mammals (Takahashi et al. 1994; Lowrey and Takahashi 2011). Our initial discovery of the *Clock* gene using forward genetic screens and positional cloning (Vitaterna et al. 1994; Antoch et al. 1997; King et al. 1997), and the identification of BMAL1 as the heterodimeric partner of CLOCK (Gekakis et al. 1998), led to idea that the CLOCK:BMAL1 transcriptional activator complex

J.S. Takahashi (✉)
Howard Hughes Medical Institute, University of Texas Southwestern Medical Center, Dallas, TX 75390, USA

Department of Neuroscience, University of Texas Southwestern Medical Center, Dallas, TX 75390, USA
e-mail: joseph.takahashi@utsouthwestern.edu

P. Sassone-Corsi, Y. Christen (eds.), *A Time for Metabolism and Hormones*,
Research and Perspectives in Endocrine Interactions,
DOI 10.1007/978-3-319-27069-2_2

13

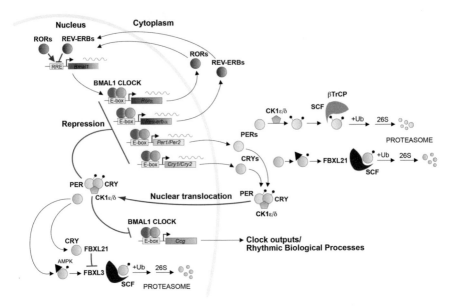

Fig. 1 Model of the circadian clock in mammals. CLOCK and BMAL1 act as master transcription factors to regulate: (1) the *Per* and *Cry* genes in the core feedback loop of the clock; (2) the REV-ERB/ROR feedback loop regulating *Bmal1* transcription; and (3) thousands of target genes that are clock outputs. The stability of the PER and CRY proteins is tightly regulated by E3 ubiquitin ligases in both the cytoplasm and nucleus that determine circadian period (Adapted from Mohawk et al. 2012 and Yoo et al. 2013)

was upstream of the *Period* and *Cryptochrome* genes, whose gene products then repressed CLOCK:BMAL1 to form an autoregulatory transcriptional feedback loop (Lowrey and Takahashi 2000). Since the identification of these "core circadian clock genes" (i.e., *Clock, Bmal1, Per1, Per2, Cry1* and *Cry2*), additional feedback loops driven by CLOCK:BMAL1, such as the loop involving *Rev-erbα* to repress *Bmal1* transcription, have been described (Preitner et al. 2002). In addition, the regulation of the stability of the PER and CRY proteins by specific E3 ubiquitin ligase complexes has been found to be important for determining the periodicity of the circadian oscillation (Busino et al. 2007; Gallego and Virshup 2007; Siepka et al. 2007; Meng et al. 2008; Yoo et al. 2013). Together, this work has led to a description of a model of the circadian clock in mammals (Fig. 1).

With the discovery and cloning of clock genes came the realization that their expression was ubiquitous (Lowrey and Takahashi 2004). We now accept that clock genes are housekeeping genes and are expressed in essentially all cells. What was perhaps even more surprising was the observation made using circadian gene reporter technology that essentially every peripheral organ system and tissue has the capacity to express autonomous circadian rhythms (Yoo et al. 2004). Thus the ubiquitous expression of clock genes is a reflection of the ubiquitous capacity of most tissues and cells to express circadian oscillations. These distributed circadian oscillators are cell autonomous and can function independently of the central clock located in the suprachiasmatic nucleus (SCN) (Nagoshi et al. 2004; Welsh et al. 2004; Yoo et al. 2004). The realization that the body is composed of a multitude of cell-autonomous clocks has raised a number of questions concerning the organization of the clock system and the role of the SCN clock in "circadian organization." Elsewhere, we have also explored the role of the SCN as a master pacemaker to synchronize peripheral oscillators (Yoo et al. 2004; Hong et al. 2007; Kornmann et al. 2007; Buhr et al. 2010; Hughes et al. 2012), as well as the role of intercellular coupling in the robustness of the SCN oscillator (Liu et al. 2007; Buhr et al. 2010; Ko et al. 2010; Welsh et al. 2010).

Structural Biology of Clock Proteins

Despite our general knowledge of clock components and their interactions, the biochemical mechanisms of circadian clock proteins and how they function within the circadian feedback loop are largely unknown. For example, many coding mutations have been described for mammalian clock proteins but, at a macroscopic level, we have little hope of understanding how they exert their phenotypic effects without a deeper understanding of their molecular mechanism. For these reasons, we have turned to structural biology to understand circadian proteins at an atomic level of resolution. Recently, we have solved the three-dimensional structure of the CLOCK:BMAL1 heterodimeric transcriptional activator complex using X-ray crystallography (Huang et al. 2012). The CLOCK:BMAL1 structure reveals an asymmetric heterodimer in which the bHLH, PAS-A and PAS-B domains of each subunit interact with their complementary domains but do so in an unexpected manner (Fig. 2). The PAS-A domains dimerize via symmetrical interactions involving α-helical domains (that are N-terminal to the canonical PAS fold) that pack against the β-sheet surfaces of the PAS-A domains (Fig. 3a). In contrast, the PAS-B domains dimerize in an asymmetric, head-to-tail fashion so that the β-sheet surface of BMAL1 interacts with the α-helical surface of CLOCK (Fig. 3b). A conserved BMAL1 Trp427 residue on an H-I loop (connecting the Hβ and Iβ strands) inserts into a hydrophobic pocket on the α-helical surface of CLOCK that resembles the co-factor binding pocket in other PAS proteins. Interestingly, a Trp residue is also conserved on the H-I loops of CLOCK, PER1 and PER2 PAS domains, suggesting that an aromatic residue inserting into the PAS receiver pocket may represent a common motif for PAS domain interactions (Crane 2012).

The structure of CLOCK:BMAL1 represents a starting point for understanding at an atomic level the mechanism driving the mammalian circadian clock. Many of the previously identified mutations on CLOCK and BMAL1 can be mapped onto the structure and, for example, predict regions of interaction of CLOCK with the CRY proteins (Huang et al. 2012). The crystal structures for the PAS-A/PAS-B domains of the mammalian PERIOD proteins (Hennig et al. 2009; Kucera et al. 2012), for the photolyase homology domains of the mammalian CRY1 (Czarna et al. 2013) and CRY2 (Xing et al. 2013) proteins, and for the CRY2/PER2-CRY binding domain complex (Nangle et al. 2014) beg the question of how

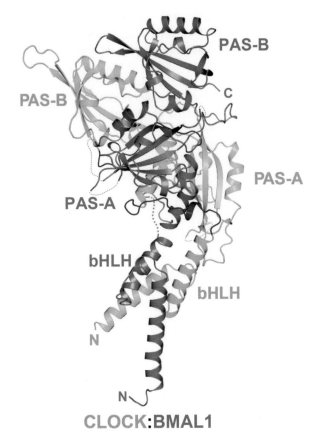

Fig. 2 CLOCK:BMAL1 structure showing bHLH, PAS-A and PAS-B domains. Linker regions shown in *red* or *orange* (From Huang et al. 2012)

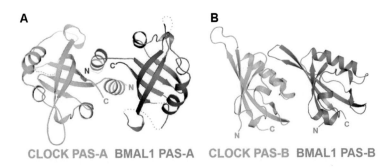

Fig. 3 PAS domains of CLOCK:BMAL1. (**a**) PAS-A interactions shown looking down the axis of the complex. (**b**) PAS-B interactions shown from a *side view* (From Huang et al. 2012)

PER and CRY interact with CLOCK:BMAL1 to repress their function. Because the native CLOCK:BMAL1/PER:CRY quaternary complexes are megadalton in size and involve other interacting proteins, and because important domains of these proteins are flexible, the solution of these complexes likely will require a combination of crystallography, NMR, and cryo electron microscopy methods in future work.

Transcriptional Architecture and Chromatin Dynamics of the Clock

To define the *cis*-acting targets of the core circadian transcriptional regulators, we used chromatin immunoprecipitation followed by sequencing (ChIP-seq) to locate DNA binding sites for BMAL1, CLOCK, NPAS2, PER1, PER2, CRY1 and CRY2 in vivo in murine liver at six times during the circadian cycle. Figure 4 shows a browser view of the *Dbp* locus, a major target gene of CLOCK-BMAL1 (Ripperger and Schibler 2006). The activators BMAL1, CLOCK and NPAS2 bind in a cyclic manner between CT0 and CT12 (CT = circadian time; CT0 is the beginning of the subjective day; CT12 is the beginning of the subjective night) at three locations in the promoter, intron 1 and intron 2. PER1, PER2 and CRY2 bind the same sites with an opposite phase at CT12-20. CRY1 exhibits a third pattern that peaks at CT0.

In genome-wide analysis, CLOCK and BMAL1 bind to over 4600 and 5900 sites, respectively, corresponding to ~3000 unique genes (Koike et al. 2012). The repressors CRY1 and CRY2 bind to significantly more sites, and many thousands of these sites are independent of CLOCK:BMAL1 and reveal DNA binding motifs for nuclear receptors (Koike et al. 2012), including the glucocorticoid receptor consistent with recent work (Lamia et al. 2011). To examine functional readouts, we used whole transcriptome RNA-seq to profile cycling genes in the liver using samples taken every 4 h over 48 h (Koike et al. 2012). Using the intron RNA signal as a proxy for pre-mRNA, we found ~1300 cycling genes and, surprisingly, they were clustered in time with a peak at CT15 (Fig. 5). To explore the possible origins of the global rhythms in nascent transcription, we analyzed the genome-wide occupancy of RNA polymerase II (RNAPII) as a function of the circadian cycle. The large subunit of RNAPII contains a C-terminal domain (CTD) that is modified at various stages of transcription (Sims et al. 2004; Fuda et al. 2009). RNAPII is recruited into the pre-initiation complex with a hypophosphorylated CTD that is recognized by the 8WG16 antibody (Jones et al. 2004). Again to our surprise, we found that RNAPII-8WG16 occupancy was highly circadian across the genome in the liver, with a peak at CT14.5, which preceded the intron RNA peak by 0.5 h (Fig. 5). Initiation of RNAPII involve phosphorylation on serine 5 (Ser5P) on the CTD of RNAPII and is recognized by the 3E8 antibody (Chapman et al. 2007). We found that RNAPII-Ser5P occupancy was also circadian, with over 13,000 sites that were significant for cycling. The timing of RNAPII-Ser5P peaked at CT0 and coincided with the peak of CRY1. At this time we found an association of CRY1, CLOCK,

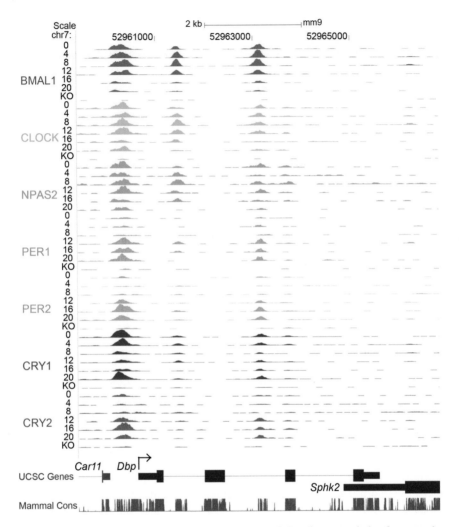

Fig. 4 UCSC genome browser view of ChIP-seq profiles of circadian transcription factors at the *Dbp* gene at six circadian times of day. BMAL1 (*blue*), CLOCK (*green*), NPAS2 (*dark green*), PER1 (*orange*), PER2 (*gold*), CRY1 (*red*), CRY2 (*pink*). 0, 4, 8, 12, 16, 20 CT (h). *KO* knockout. (From Koike et al. 2012)

BMAL1 and RNAPII-Ser5P binding sites, suggesting that CLOCK:BMAL1 could recruit and initiate RNAPII but CRY1 repressed the complex leading to a "poised" state.

Given the genome-wide circadian rhythms of RNAPII occupancy, we assessed chromatin states associated with transcription initiation and elongation during the circadian cycle. Figure 6 shows a browser view of six histone modifications that are characteristic of promoters, enhancers and transcription elongation (Kim et al. 2005; Barski et al. 2007; Guenther et al. 2007; Li et al. 2007; Creyghton

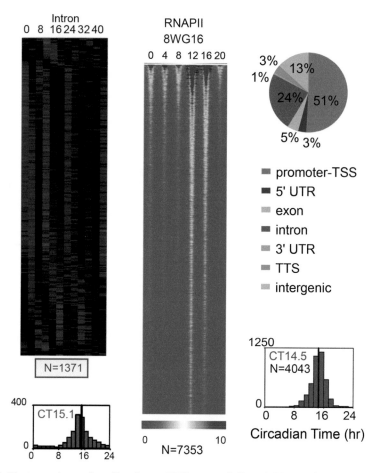

Fig. 5 Heatmap views of cycling intron RNA genes (*left*) and RNAPII-8WG16 occupancy (*right*). More than 4000 peaks had significant circadian RNAPII binding (From Koike et al. 2012)

et al. 2010; Ong and Corces 2011; Rada-Iglesias et al. 2011). Histone H3K4me3, H3K9ac and H3K27ac are enriched at promoters and show robust circadian rhythms in occupancy at the *Dbp* gene. When examined across the genome, we found that circadian rhythms in RNAPII occupancy as well as histone H3K4me3, H3K9ac and H3K27ac modifications occurred in the majority of expressed genes, even in cases where cycling RNA could not be detected. Thus a third surprise in this work was the observation that chromatin states were being modulated in a circadian manner across the genome in the liver.

What accounts for these genome-wide circadian rhythms in RNAPII occupancy and histone modifications? Examination of the relationship between circadian transcription factor occupancy and gene expression shows that approximately 90 % of genes bound by these factors are expressed whereas only 1–5 % of unexpressed genes are similarly bound (Koike et al. 2012). These results

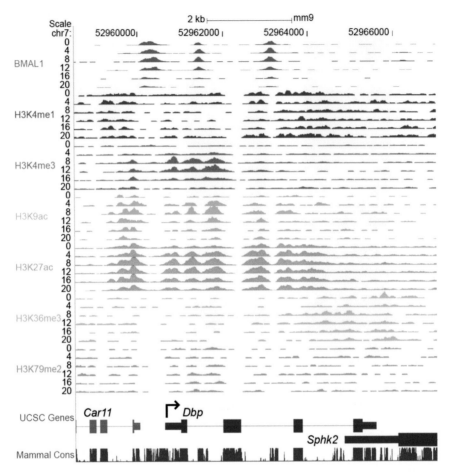

Fig. 6 UCSC genome browser view of histone methylation and acetylation at the *Dbp* gene. BMAL1 (*blue*), H3K4me1 (*red*), H3K4me3 (*pink*), H3K9ac (*aqua*), H3K27ac (*orange*), H3K36me3 (*green*), H3K79me2 (*dark green*) (From Koike et al. 2012)

demonstrate that gene expression per se rather than rhythmic gene expression is tightly correlated with circadian transcription factor binding. Rhythmic circadian transcription factor occupancy in turn could then be responsible for RNAPII recruitment and initiation on a genome-wide basis, which would then lead to the global rhythmic histone modifications seen here. Thus, circadian transcriptional regulators appear to be involved in the initial stages of RNAPII recruitment and initiation and the histone modifications associated with these events to set the stage for gene expression on a global scale, but additional control steps must then determine the ultimate transcriptional outputs from these sites.

In summary, we have defined the *cis*-regulatory network of the entire core circadian transcriptional regulatory loop on a genome scale and found a highly stereotyped, time-dependent pattern of core transcription factor binding, RNAPII

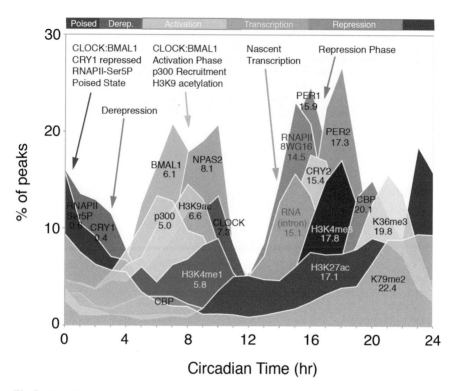

Fig. 7 Circadian transcriptional landscape in the liver. Histograms show the phase distributions of each factor as a function of time of day. *Derep* derepression (From Koike et al. 2012)

occupancy, RNA expression and chromatin states (Fig. 7). We defined three distinctive phases of the circadian cycle: (1) a poised phase in which CLOCK: BMAL1 and CRY1 bind to E-box sites in a transcriptionally silent state associated with RNAPII-Ser5P; (2) a temporally coordinated transcriptional activation phase in which RNAPII and p300 recruitment, pre-mRNA transcript expression, and H3K9ac, H3K4me3 and H3K27ac occupancy oscillate; and (3) a repression phase in which PER1, PER2 and CRY2 occupancy peaks. Circadian modulation of RNAPII recruitment and chromatin remodeling occurs on a genome-wide scale far greater than that seen previously by gene expression profiling. Thus, the circadian clock in the liver modulates the occupancy of RNAPII across the genome, leading at least in part to genome-wide circadian modulation of chromatin states that, in turn, poise the genome for transcription on a daily basis to act in concert with the daily metabolic demands of the organism.

Acknowledgment J.S.T. is an Investigator in the Howard Hughes Medical Institute. Research also supported by awards from the NIH.

References

Antoch MP, Song EJ, Chang AM, Vitaterna MH, Zhao Y, Wilsbacher LD, Sangoram AM, King DP, Pinto LH, Takahashi JS (1997) Functional identification of the mouse circadian Clock gene by transgenic BAC rescue. Cell 89:655–667

Barski A, Cuddapah S, Cui K, Roh TY, Schones DE, Wang Z, Wei G, Chepelev I, Zhao K (2007) High-resolution profiling of histone methylations in the human genome. Cell 129:823–837

Buhr ED, Yoo SH, Takahashi JS (2010) Temperature as a universal resetting cue for mammalian circadian oscillators. Science 330:379–385

Busino L, Bassermann F, Maiolica A, Lee C, Nolan PM, Godinho SI, Draetta GF, Pagano M (2007) SCF^{Fbxl3} controls the oscillation of the circadian clock by directing the degradation of cryptochrome proteins. Science 316:900–904

Chapman RD, Heidemann M, Albert TK, Mailhammer R, Flatley A, Meisterernst M, Kremmer E, Eick D (2007) Transcribing RNA polymerase II is phosphorylated at CTD residue serine-7. Science 318:1780–1782

Crane BR (2012) Biochemistry. Nature's intricate clockwork. Science 337:165–166

Creyghton MP, Cheng AW, Welstead GG, Kooistra T, Carey BW, Steine EJ, Hanna J, Lodato MA, Frampton GM, Sharp PA, Boyer LA, Young RA, Jaenisch R (2010) Histone H3K27ac separates active from poised enhancers and predicts developmental state. Proc Natl Acad Sci USA 107:21931–21936

Czarna A, Berndt A, Singh H, Grudziecki A, Ladurner A, Timinszky G, Kramer A, Wolf E (2013) Structures of Drosophila cryptochrome and mouse cryptochrome1 provide insight into circadian function. Cell 153:1394–1405

Fuda N, Ardehali M, Lis J (2009) Defining mechanisms that regulate RNA polymerase II transcription in vivo. Nature 461:186–192

Gallego M, Virshup DM (2007) Post-translational modifications regulate the ticking of the circadian clock. Nat Rev Mol Cell Biol 8:139–148

Gekakis N, Staknis D, Nguyen HB, Davis FC, Wilsbacher LD, King DP, Takahashi JS, Weitz CJ (1998) Role of the CLOCK protein in the mammalian circadian mechanism. Science 280:1564–1569

Guenther MG, Levine SS, Boyer LA, Jaenisch R, Young RA (2007) A chromatin landmark and transcription initiation at most promoters in human cells. Cell 130:77–88

Hennig S, Strauss H, Vanselow K, Yildiz O, Schulze S, Arens J, Kramer A, Wolf E (2009) Structural and functional analyses of PAS domain interactions of the clock proteins Drosophila PERIOD and mouse PERIOD2. PLoS Biol 7:e94

Hong HK, Chong JL, Song W, Song EJ, Jyawook AA, Schook AC, Ko CH, Takahashi JS (2007) Inducible and reversible Clock gene expression in brain using the tTA system for the study of circadian behavior. PLoS Genet 3:e33

Huang N, Chelliah Y, Shan Y, Taylor CA, Yoo SH, Partch C, Green CB, Zhang H, Takahashi JS (2012) Crystal structure of the heterodimeric CLOCK:BMAL1 transcriptional activator complex. Science 337:189–194

Hughes ME, Hong HK, Chong JL, Indacochea AA, Lee SS, Han M, Takahashi JS, Hogenesch JB (2012) Brain-specific rescue of Clock reveals system-driven transcriptional rhythms in peripheral tissue. PLoS Genet 8:e1002835

Jones JC, Phatnani HP, Haystead TA, MacDonald JA, Alam SM, Greenleaf AL (2004) C-terminal repeat domain kinase I phosphorylates Ser2 and Ser5 of RNA polymerase II C-terminal domain repeats. J Biol Chem 279:24957–24964

Kim TH, Barrera LO, Zheng M, Qu C, Singer MA, Richmond TA, Wu Y, Green RD, Ren B (2005) A high-resolution map of active promoters in the human genome. Nature 436:876–880

King DP, Zhao Y, Sangoram AM, Wilsbacher LD, Tanaka M, Antoch MP, Steeves TD, Vitaterna MH, Kornhauser JM, Lowrey PL, Turek FW, Takahashi JS (1997) Positional cloning of the mouse circadian clock gene. Cell 89:641–653

Ko CH, Yamada YR, Welsh DK, Buhr ED, Liu AC, Zhang EE, Ralph MR, Kay SA, Forger DB, Takahashi JS (2010) Emergence of noise-induced oscillations in the central circadian pacemaker. PLoS Biol 8:e1000513

Koike N, Yoo SH, Huang HC, Kumar V, Lee C, Kim TK, Takahashi JS (2012) Transcriptional architecture and chromatin landscape of the core circadian clock in mammals. Science 338:349–354

Kornmann B, Schaad O, Bujard H, Takahashi JS, Schibler U (2007) System-driven and oscillator-dependent circadian transcription in mice with a conditionally active liver clock. PLoS Biol 5:e34

Kucera N, Schmalen I, Hennig S, Öllinger R, Strauss H, Grudziecki A, Wieczorek C, Kramer A, Wolf E (2012) Unwinding the differences of the mammalian PERIOD clock proteins from crystal structure to cellular function. Proc Natl Acad Sci USA 109:3311–3316

Lamia K, Papp S, Yu R, Barish G, Uhlenhaut N, Jonker J, Downes M, Evans R (2011) Cryptochromes mediate rhythmic repression of the glucocorticoid receptor. Nature 480:552–556

Li B, Carey M, Workman JL (2007) The role of chromatin during transcription. Cell 128:707–719

Liu AC, Welsh DK, Ko CH, Tran HG, Zhang EE, Priest AA, Buhr ED, Singer O, Meeker K, Verma IM, Doyle FJ 3rd, Takahashi JS, Kay SA (2007) Intercellular coupling confers robustness against mutations in the SCN circadian clock network. Cell 129:605–616

Lowrey PL, Takahashi JS (2000) Genetics of the mammalian circadian system: photic entrainment, circadian pacemaker mechanisms, and posttranslational regulation. Annu Rev Genet 34:533–562

Lowrey PL, Takahashi JS (2004) Mammalian circadian biology: elucidating genome-wide levels of temporal organization. Annu Rev Genomics Hum Genet 5:407–441

Lowrey PL, Takahashi JS (2011) Genetics of circadian rhythms in mammalian model organisms. Adv Genet 74:175–230

Meng QJ, Logunova L, Maywood ES, Gallego M, Lebiecki J, Brown TM, Sladek M, Semikhodskii AS, Glossop NR, Piggins HD, Chesham JE, Bechtold DA, Yoo SH, Takahashi JS, Virshup DM, Boot-Handford RP, Hastings MH, Loudon AS (2008) Setting clock speed in mammals: the CK1 epsilon tau mutation in mice accelerates circadian pacemakers by selectively destabilizing PERIOD proteins. Neuron 58:78–88

Mohawk JA, Green CB, Takahashi JS (2012) Central and peripheral circadian clocks in mammals. Annu Rev Neurosci 35:445–462

Nagoshi E, Saini C, Bauer C, Laroche T, Naef F, Schibler U (2004) Circadian gene expression in individual fibroblasts; cell-autonomous and self-sustained oscillators pass time to daughter cells. Cell 119:693–705

Nangle SN, Rosensweig C, Koike N, Tei H, Takahashi JS, Green CB, Zheng N (2014) Molecular assembly of the period-cryptochrome circadian transcriptional repressor complex. Elife 3:e03674

Ong CT, Corces VG (2011) Enhancer function: new insights into the regulation of tissue-specific gene expression. Nat Rev Genet 12:283–293

Preitner N, Damiola F, Lopez-Molina L, Zakany J, Duboule D, Albrecht U, Schibler U (2002) The orphan nuclear receptor REV-ERBα controls circadian transcription within the positive limb of the mammalian circadian oscillator. Cell 110:251–260

Rada-Iglesias A, Bajpai R, Swigut T, Brugmann SA, Flynn RA, Wysocka J (2011) A unique chromatin signature uncovers early developmental enhancers in humans. Nature 470:279–283

Ripperger JA, Schibler U (2006) Rhythmic CLOCK-BMAL1 binding to multiple E-box motifs drives circadian Dbp transcription and chromatin transitions. Nat Genet 38:369–374

Siepka SM, Yoo SH, Park J, Song W, Kumar V, Hu Y, Lee C, Takahashi JS (2007) Circadian mutant overtime reveals F-box protein FBXL3 regulation of cryptochrome and period gene expression. Cell 129:1011–1023

Sims RJ 3rd, Belotserkovskaya R, Reinberg D (2004) Elongation by RNA polymerase II: the short and long of it. Genes Dev 18:2437–2468

Takahashi JS, Pinto LH, Vitaterna MH (1994) Forward and reverse genetic approaches to behavior in the mouse. Science 264:1724–1733

Vitaterna MH, King DP, Chang AM, Kornhauser JM, Lowrey PL, McDonald JD, Dove WF, Pinto LH, Turek FW, Takahashi JS (1994) Mutagenesis and mapping of a mouse gene, Clock, essential for circadian behavior. Science 264:719–725

Welsh DK, Yoo SH, Liu AC, Takahashi JS, Kay SA (2004) Bioluminescence imaging of individual fibroblasts reveals persistent, independently phased circadian rhythms of clock gene expression. Curr Biol 14:2289–2295

Welsh DK, Takahashi JS, Kay SA (2010) Suprachiasmatic nucleus: cell autonomy and network properties. Annu Rev Physiol 72:551–577

Xing W, Busino L, Hinds T, Marionni S, Saifee N, Bush M, Pagano M, Zheng N (2013) SCFFBXL3 ubiquitin ligase targets cryptochromes at their cofactor pocket. Nature 496:64–68

Yoo SH, Yamazaki S, Lowrey PL, Shimomura K, Ko CH, Buhr ED, Siepka SM, Hong HK, Oh WJ, Yoo OJ, Menaker M, Takahashi JS (2004) PERIOD2::LUCIFERASE real-time reporting of circadian dynamics reveals persistent circadian oscillations in mouse peripheral tissues. Proc Natl Acad Sci USA 101:5339–5346

Yoo SH, Mohawk JA, Siepka SM, Shan Y, Huh SK, Hong HK, Kornblum I, Kumar V, Koike N, Xu M, Nussbaum J, Liu X, Chen Z, Chen ZJ, Green CB, Takahashi JS (2013) Competing E3 ubiquitin ligases govern circadian periodicity by degradation of CRY in nucleus and cytoplasm. Cell 152:1091–1105

Circadian Mechanisms in Bioenergetics and Cell Metabolism

Joseph Bass

Abstract Circadian clocks are biologic oscillators present in all photosensitive species that produce 24-h cycles in the transcription of rate-limiting metabolic enzymes in anticipation of the light–dark cycle. In mammals, the clock drives energetic cycles to maintain physiologic constancy during the daily switch in behavioral (sleep/wake) and nutritional (fasting/feeding) states. A molecular connection between circadian clocks and tissue metabolism was first established with the discovery that 24-h transcriptional rhythms are cell-autonomous and self-sustained in cultured fibroblasts, and that clocks are present in most tissues and comprise a robust temporal network throughout the body. A central question remains: how do circadian transcriptional programs integrate physiologic systems within individual cells of the intact animal and how does the ensemble of local clocks align temporal harmonics in the organism with the environment? Our approach to studies of metabolic regulation by the molecular clock began with analyses of metabolic pathologies in circadian mutant animals, experiments that first became possible with the cloning of the clock genes in the late 1990s. A paradox in our early studies was that the effects of circadian clock disruption were both nutrient- and time-dependent, so that, under fed conditions, animals exhibited diabetes whereas during fasting, they decompensated and died. Application of a broad range of tissue-specific genetic and biochemical approaches has now begun to provide mechanistic insight into the circadian control of metabolism.

J. Bass, M.D., Ph.D. (✉)
Department of Medicine, Division of Endocrinology, Metabolism and Molecular Medicine, Northwestern University, Feinberg School of Medicine, 303 East Superior Street Lurie 7-107, Chicago, IL 60611, USA

Department of Neurobiology, Northwestern University, Evanston, IL 60208, USA
e-mail: j-bass@northwestern.edu

© The Author(s) 2016
P. Sassone-Corsi, Y. Christen (eds.), *A Time for Metabolism and Hormones*,
Research and Perspectives in Endocrine Interactions,
DOI 10.1007/978-3-319-27069-2_3

Genetic Approaches to Dissecting Circadian Physiology

Glucose homeostasis is a dynamic process that is subjected to rhythmic variation throughout the daily light–dark cycle. Impaired glucose regulation arises from desynchrony in the integration of anabolic, catabolic, and incretin hormones across the circadian cycle and leads to metabolic syndrome and diabetes mellitus, disorders that are associated with over-nutrition, sedentary lifestyle, and sleep-wake disruption common in industrialized society. Individuals with diabetes must adjust their insulin levels differently every day and night even independently of how much they eat; however, the molecular underpinnings of circadian glucose regulation were previously not well understood. Genome-wide association and deep-sequencing studies have shown that variants of the *melatonin receptor 1b* and *cryptochrome 2* genes correlate with glucose variation in humans, suggesting a genetic linkage between the circadian system and glucoregulatory processes in man (Bouatia-Naji et al. 2009; Mulder et al. 2009; Dupuis et al. 2010). Against this backdrop, work from our laboratory using circadian clock mutant mice first revealed an essential role for the intrinsic beta cell clock in insulin secretion, beta cell development, and diabetes mellitus (Marcheva et al. 2010). Subsequent studies in three other groups have corroborated our observation that local function of the clock transcription factor in islets is crucial for normal glucose homeostasis (Sadacca et al. 2011; Lee et al. 2013; Pulimeno et al. 2013). Importantly, work from the Dibner laboratory has been the first to manipulate and monitor clock function in isolated human islet cells, raising the possibility that future investigation into circadian cell physiology will yield new understanding of beta cell failure in man (Pulimeno et al. 2013). In recently published work, we have developed tamoxifen-inducible Cre-LoxP technology to conditionally eliminate clock gene function in pancreas (PMID 26542580). Remarkably, our results establish that acute pancreatic clock ablation in the adult is sufficient to cause diabetes mellitus in the whole animal. These new genetic studies are the first to demonstrate an essential role for the adult circadian system in beta cell glucose regulation, although a gap remains in our understanding of the cell and molecular bases for clock function in the beta cell. Using conditional gene targeting and next-generation sequencing described in the following sections, we are presently poised to dissect the genomic, biochemical, and physiologic mechanism of the clock in beta cell failure. Moreover, since clock transcription factors impact both islet cell growth and stress response, we also seek to understand the role of the islet cell clock in susceptibility to beta cell apoptosis in type 1 diabetes, in islet regenerative capacity, and in islet cell survival in insulin resistant obesity.

Our analysis of the beta cell clock also opens broader insight into the role of transcription factor deregulation in beta cell failure and the unifying and distinct molecular events between tissues that culminate in diabetes mellitus. In this regard, positional cloning of genes causing Maturity Onset Diabetes of the Young (MODY) in humans has revealed that the hepatic nuclear factor (HNF) network of forkhead transcription factors plays a critical role in beta cell development and function, although knowledge of the cell-context specific determinants of HNF action in liver and pancreas remains incomplete. By analogy, an important goal in understanding how beta cell transcription defects related to the clock pathway give rise to diabetes will be to elucidate differences in the clock-controlled enhancer network in liver and pancreas and to compare the cistromes and transcriptomes for these factors in each tissue. Interestingly, there is coincidence of CLOCK/BMAL1 binding sites in liver at loci marked by monomethylated H3K4 in pancreas, although it is not known whether the enhancer state (poised, H3K4me1 vs active, H3K4me1 with H3K27Ac) in pancreas varies over the 24-h cycle. In the long term it will be necessary to evaluate the cistromes and localization of clock factors with established transcription factors involved in beta cell function over the full circadian cycle in both liver and pancreas. Such studies will further elucidate the mechanism by which the clock controls gene transcription networks involved in insulin release, namely by determining the extent to which CLOCK/BMAL1 directly binds to promoters and/or enhancers or regulates epigenetic chromatin modifiers that determine accessibility to transcription factors and RNA polymerase genome-wide. Overall, studies of the beta cell molecular clock will elucidate how glucose homeostasis is coupled to the light/dark cycle and the transcriptional determinants of circadian physiology.

Clock-NAD$^+$-Sirtuin Pathway in Bioenergetics

A major step in understanding how the clock-NAD$^+$ cycle impacts physiology came from the observations that circadian mutant mice become hypoglycemic and die when subject to a prolonged fast (PMID 24051248, unpublished data) and also exhibit muscle and heart failure (PMID 20956306, 21452915), all hallmarks of mitochondrial disease, which prompted us to dissect the mechanisms of clock regulation of mitochondrial function. We began our investigation into the effect of NAD$^+$ deficiency on mitochondrial function in circadian mutant animals using a multi-faceted approach, including unbiased proteomics, which led to the identification of abnormal acetylation of enzymes involved in lipid oxidation, amino acid catabolism, tricarboxylic acid (TCA) cycle, electron transport chain (ETC), and superoxide dismutase pathways. Importantly, loss-of-function mutations in several of these oxidative enzymes have also been identified in the human metabolic myopathy syndrome and in both glioblastoma and renal cell carcinoma, indicating a broader effect of the clock-NAD$^+$ pathway on mitochondrial metabolism in both normal and transformed cells. Using tissue- and cell-based bioenergetics assays, we discovered that abrogation of the clock impairs electron transfer from lipid to the TCA cycle, in addition to increased mitochondrial production of superoxide free radical, increasing sensitivity to genotoxic stress. Our work also showed that cells exhibit an autonomous rhythm of oxygen consumption, glucose oxidation, and mitochondrial lipid catabolism. Importantly, the oxygen consumption cycle in muscle is directly linked to metabolism of NAD$^+$ and activity of the mitochondrial NAD$^+$-dependent deacetylase SIRT3 (Peek et al. 2013).

Although the aforementioned work has pinpointed specific defects in clock control of mitochondrial function, several unanswered questions remain in dissecting the effect of clock-NAD$^+$ rhythms on physiology and cell biology. First, it is not yet known how NAD$^+$ deficiency locally within skeletal muscle contributes to respiration or exercise tolerance in circadian mutant mice or in animals subjected to environmental circadian disruption. Though skeletal muscle ablation of the clock has been achieved in our group and others, the biochemical pathways through which clock abrogation impairs oxidative capacity remain largely unknown (Dyar et al. 2014).

Second, we still do not know whether clock abrogation and NAD$^+$ deficiency in liver or skeletal muscle impacts overall energy balance and alters the capacity to utilize carbohydrate and lipid as a fuel source. New pharmacologic (Wang et al. 2014) and genetic means to raise NAD$^+$ both globally in the whole animal and selectively within either liver or skeletal muscle are now available and will be powerful tools in evaluating the potential to boost NAD$^+$ as a therapeutic strategy in myopathy and liver defects of circadian mutant animals. Finally, in addition to its function as a cofactor for the class III histone deacetylases, NAD$^+$ is a cofactor for the poly-ADP-ribosylases, critical factors in DNA repair and stress response, though the possible interaction between rhythmic regulation of NAD$^+$ and PARP activity is not known. Lastly, NAD$^+$ functions as an electron transport molecule and, as such, it is a direct marker of cellular redox state and the balance between glycolytic and oxidative metabolism. Whether NAD$^+$ might participate in the bidirectional communication between metabolism and the clock system remains an area of intensive investigation. In summary, discovery of the clock as an upstream regulator of NAD$^+$ provides a wealth of opportunity to dissect the interrelationship between circadian rhythms, physiology, and epigenetics.

Reciprocal Control of the Clock by Nutrient

Circadian clocks are biologic oscillators that produce 24-h cycles in the transcription of rate-limiting metabolic enzymes in anticipation of the solar cycle. The molecular clock is programmed by a transcription-translation feedback loop that is comprised of activators (CLOCK/BMAL1) that induce the expression of their own repressors (CRYs/PERs) in a cycle that repeats itself every 24 h. The REV-ERB and ROR proteins form an ancillary loop that modulates *Bmal1* transcription. In animals, clocks are organized hierarchically, with brain pacemaker cells synchronizing peripheral tissue clocks, leading to a classical view of the central clock as the main driver of metabolism. However, circadian oscillations within both brain and peripheral tissues have recently been shown to be sensitive to timing of nutrient availability and can become uncoupled from the light–dark cycle, as demonstrated by experimentally restricting food access to the light cycle when mice are normally resting (Damiola et al. 2000; Stokkan et al. 2001). Further, simply substituting regular with high fat chow in mice fed ad libitum lengthens periodicity of locomotor activity and alters peripheral metabolic rhythms, providing further evidence for a bidirectional relationship between clock function and metabolism (Kohsaka et al. 2007). Our discovery that diet-induced obesity reprograms both the cellular molecular clock and behavior revealed for the first time that a controlled change in nutritional environment leads to altered circadian rhythms. This idea, that circadian and metabolic systems reciprocally interact and that perturbation of the metabolic environment alters the homeostatic relationship between these systems, has been widely confirmed but still remains poorly understood at the mechanistic level. Human analyses, including genome-wide association studies, population based case–control investigation, and clinical research, have cumulatively indicated a strong interrelationship between circadian disruption, obesity, diabetes mellitus, and metabolic syndrome. Moreover, certain inflammatory and cardiovascular events, including thrombosis and nocturnal asthma, exhibit pronounced circadian variation. Surprisingly, dietary macronutrient directly impacts behavioral and molecular clock function, and circadian disruption itself exacerbates the progression of diet-induced obesity, exerting distinct effects within local metabolic organs. Moreover, limiting high-fat food to the incorrect circadian phase accelerates weight gain, whereas limiting high-fat feeding to the correct phase ameliorates hepatic steatosis, a hallmark of metabolic syndrome (Maury et al. 2010). While we previously demonstrated that diet-induced obesity reprograms the cellular molecular clock and circadian behavior, we have more recently sought to identify the macronutrient disruptor of circadian behavior by providing mice an isocaloric diet high in either saturated or unsaturated fats (SFD and UFD). Our goal is to identify the mechanism by which a macronutrient directly alters behavior and neuronal circadian pacemaker function. We propose that nutrient signaling plays a central role in inter-organ circadian communication and that circadian disruption induced by high saturated fat contributes to the rate of progression of metabolic syndrome.

Summary and Future Directions

A major window to understanding how the clock is coupled to metabolism was opened with discovery of metabolic syndrome pathologies in multi-tissue circadian mutant mice, including susceptibility to diet-induced obesity, mis-timed feeding rhythms, hypoinsulinemia, and energetic collapse upon fasting. Using Cre-LoxP conditional transgenesis and dynamic endocrine testing, we have pinpointed the tissue-specific role of the clock in energy and glucose homeostasis, with our most detailed understanding of this process in liver, muscle, and endocrine pancreas. In the post-prandial condition, the beta cell clock is essential for nutrient and adenyl cyclase-induced insulin exocytosis. In contrast, the hepatocyte and myocyte clocks are required for oxidative metabolism. Circadian mutant mice die upon prolonged fasting due to mitochondrial failure, a defect that we have tied to the bioavailability of NAD^+, a cofactor of the class III histone deacetylases and poly-ADP ribosylase enzymes involved in adjusting metabolic and gene regulation in response to environmental change, including glucose deprivation, oxidative damage, and cell stress. Indeed, we have found that liver and myoblasts exhibit an autonomous rhythm of oxygen consumption, glucose oxidation, and mitochondrial lipid catabolism that is directly linked to an autonomous rhythm of NAD^+ metabolism and, consequently, to cyclic activity of the mitochondrial NAD^+-dependent deacetylase SIRT3. NAD^+ supplementation using the pro-drug NMN improves respiration in live animals, indicating that circadian control of NAD^+ metabolism plays a key role in cellular and organismal respiration. A future challenge will be to determine the cell and molecular basis for the interplay between nutritional and circadian processes important in metabolic health and disease states.

Acknowledgments

Funding Sources P01AG011412 (National Institute on Aging), R01DK090625 (National Institute of Diabetes and Digestive and Kidney Disease), R01DK100814 (National Institute of Diabetes and Digestive and Kidney Disease), 17-2013-511 (Juvenile Diabetes Research Foundation with Helmsley Charitable Trust), 1-INO-2014-178-A-V (Juvenile Diabetes Research Foundation with Helmsley Charitable Trust). I thank the lab members who participated in our cited research and to KM Ramsey for helpful comments and suggestions on the manuscript.

References

Bouatia-Naji N, Bonnefond A, Cavalcanti-Proenca C, Sparso T, Holmkvist J, Marchand M, Delplanque J, Lobbens S, Rocheleau G, Durand E, De Graeve F, Chevre JC, Borch-Johnsen K, Hartikainen AL, Ruokonen A, Tichet J, Marre M, Weill J, Heude B, Tauber M, Lemaire K, Schuit F, Elliott P, Jorgensen T, Charpentier G, Hadjadj S, Cauchi S, Vaxillaire M, Sladek R, Visvikis-Siest S, Balkau B, Levy-Marchal C, Pattou F, Meyre D, Blakemore AI, Jarvelin MR, Walley AJ, Hansen T, Dina C, Pedersen O, Froguel P (2009) A variant near MTNR1B is associated with increased fasting plasma glucose levels and type 2 diabetes risk. Nat Genet 41:89–94
Damiola F, Le Minh N, Preitner N, Kornmann B, Fleury-Olela F, Schibler U (2000) Restricted feeding uncouples circadian oscillators in peripheral tissues from the central pacemaker in the suprachiasmatic nucleus. Genes Dev 14:2950–2961
Dupuis J, Langenberg C, Prokopenko I, Saxena R, Soranzo N, Jackson AU, Wheeler E, Glazer NL, Bouatia-Naji N, Gloyn AL, Lindgren CM, Magi R, Morris AP, Randall J, Johnson T, Elliott P,

Rybin D, Thorleifsson G, Steinthorsdottir V, Henneman P, Grallert H, Dehghan A, Hottenga JJ, Franklin CS, Navarro P, Song K, Goel A, Perry JR, Egan JM, Lajunen T, Grarup N, Sparso T, Doney A, Voight BF, Stringham HM, Li M, Kanoni S, Shrader P, Cavalcanti-Proenca C, Kumari M, Qi L, Timpson NJ, Gieger C, Zabena C, Rocheleau G, Ingelsson E, An P, O'Connell J, Luan J, Elliott A, McCarroll SA, Payne F, Roccasecca RM, Pattou F, Sethupathy P, Ardlie K, Ariyurek Y, Balkau B, Barter P, Beilby JP, Ben-Shlomo Y, Benediktsson R, Bennett AJ, Bergmann S, Bochud M, Boerwinkle E, Bonnefond A, Bonnycastle LL, Borch-Johnsen K, Bottcher Y, Brunner E, Bumpstead SJ, Charpentier G, Chen YD, Chines P, Clarke R, Coin LJ, Cooper MN, Cornelis M, Crawford G, Crisponi L, Day IN, de Geus EJ, Delplanque J, Dina C, Erdos MR, Fedson AC, Fischer-Rosinsky A, Forouhi NG, Fox CS, Frants R, Franzosi MG, Galan P, Goodarzi MO, Graessler J, Groves CJ, Grundy S, Gwilliam R, Gyllensten U, Hadjadj S, Hallmans G, Hammond N, Han X, Hartikainen AL, Hassanali N, Hayward C, Heath SC, Hercberg S, Herder C, Hicks AA, Hillman DR, Hingorani AD, Hofman A, Hui J, Hung J, Isomaa B, Johnson PR, Jorgensen T, Jula A, Kaakinen M, Kaprio J, Kesaniemi YA, Kivimaki M, Knight B, Koskinen S, Kovacs P, Kyvik KO, Lathrop GM, Lawlor DA, Le Bacquer O, Lecoeur C, Li Y, Lyssenko V, Mahley R, Mangino M, Manning AK, Martinez-Larrad MT, McAteer JB, McCulloch LJ, McPherson R, Meisinger C, Melzer D, Meyre D, Mitchell BD, Morken MA, Mukherjee S, Naitza S, Narisu N, Neville MJ, Oostra BA, Orru M, Pakyz R, Palmer CN, Paolisso G, Pattaro C, Pearson D, Peden JF, Pedersen NL, Perola M, Pfeiffer AF, Pichler I, Polasek O, Posthuma D, Potter SC, Pouta A, Province MA, Psaty BM, Rathmann W, Rayner NW, Rice K, Ripatti S, Rivadeneira F, Roden M, Rolandsson O, Sandbaek A, Sandhu M, Sanna S, Sayer AA, Scheet P, Scott LJ, Seedorf U, Sharp SJ, Shields B, Sigurethsson G, Sijbrands EJ, Silveira A, Simpson L, Singleton A, Smith NL, Sovio U, Swift A, Syddall H, Syvanen AC, Tanaka T, Thorand B, Tichet J, Tonjes A, Tuomi T, Uitterlinden AG, van Dijk KW, van Hoek M, Varma D, Visvikis-Siest S, Vitart V, Vogelzangs N, Waeber G, Wagner PJ, Walley A, Walters GB, Ward KL, Watkins H, Weedon MN, Wild SH, Willemsen G, Witteman JC, Yarnell JW, Zeggini E, Zelenika D, Zethelius B, Zhai G, Zhao JH, Zillikens MC, Borecki IB, Loos RJ, Meneton P, Magnusson PK, Nathan DM, Williams GH, Hattersley AT, Silander K, Salomaa V, Smith GD, Bornstein SR, Schwarz P, Spranger J, Karpe F, Shuldiner AR, Cooper C, Dedoussis GV, Serrano-Rios M, Morris AD, Lind L, Palmer LJ, Hu FB, Franks PW, Ebrahim S, Marmot M, Kao WH, Pankow JS, Sampson MJ, Kuusisto J, Laakso M, Hansen T, Pedersen O, Pramstaller PP, Wichmann HE, Illig T, Rudan I, Wright AF, Stumvoll M, Campbell H, Wilson JF, Bergman RN, Buchanan TA, Collins FS, Mohlke KL, Tuomilehto J, Valle TT, Altshuler D, Rotter JI, Siscovick DS, Penninx BW, Boomsma DI, Deloukas P, Spector TD, Frayling TM, Ferrucci L, Kong A, Thorsteinsdottir U, Stefansson K, van Duijn CM, Aulchenko YS, Cao A, Scuteri A, Schlessinger D, Uda M, Ruokonen A, Jarvelin MR, Waterworth DM, Vollenweider P, Peltonen L, Mooser V, Abecasis GR, Wareham NJ, Sladek R, Froguel P, Watanabe RM, Meigs JB, Groop L, Boehnke M, McCarthy MI, Florez JC, Barroso I (2010) New genetic loci implicated in fasting glucose homeostasis and their impact on type 2 diabetes risk. Nat Genet 42:105–116

Dyar KA, Ciciliot S, Wright LE, Bienso RS, Tagliazucchi GM, Patel VR, Forcato M, Paz MI, Gudiksen A, Solagna F, Albiero M, Moretti I, Eckel-Mahan KL, Baldi P, Sassone-Corsi P, Rizzuto R, Bicciato S, Pilegaard H, Blaauw B, Schiaffino S (2014) Muscle insulin sensitivity and glucose metabolism are controlled by the intrinsic muscle clock. Mol Metab 3:29–41

Kohsaka A, Laposky AD, Ramsey KM, Estrada C, Joshu C, Kobayashi Y, Turek FW, Bass J (2007) High-fat diet disrupts behavioral and molecular circadian rhythms in mice. Cell Metab 6:414–421

Lee J, Moulik M, Fang Z, Saha P, Zou F, Xu Y, Nelson DL, Ma K, Moore DD, Yechoor VK (2013) Bmal1 and beta-cell clock are required for adaptation to circadian disruption, and their loss of function leads to oxidative stress-induced beta-cell failure in mice. Mol Cell Biol 33:2327–2338

Marcheva B, Ramsey KM, Buhr ED, Kobayashi Y, Su H, Ko CH, Ivanova G, Omura C, Mo S, Vitaterna MH, Lopez JP, Philipson LH, Bradfield CA, Crosby SD, Jebailey L, Wang X, Takahashi JS, Bass J (2010) Disruption of the clock components CLOCK and BMAL1 leads to hypoinsulinaemia and diabetes. Nature 466:571–572

Maury E, Ramsey KM, Bass J (2010) Circadian rhythms and metabolic syndrome: from experimental genetics to human disease. Circ Res 106:447–462

Mulder H, Nagorny CL, Lyssenko V, Groop L (2009) Melatonin receptors in pancreatic islets: good morning to a novel type 2 diabetes gene. Diabetologia 52:1240–1249

Peek CB, Affinati AH, Ramsey KM, Kuo HY, Yu W, Sena LA, Ilkayeva O, Marcheva B, Kobayashi Y, Omura C, Levine DC, Bacsik DJ, Gius D, Newgard CB, Goetzman E, Chandel NS, Denu JM, Mrksich M, Bass J (2013) Circadian clock NAD+ cycle drives mitochondrial oxidative metabolism in mice. Science 342:1243417

Pulimeno P, Mannic T, Sage D, Giovannoni L, Salmon P, Lemeille S, Giry-Laterriere M, Unser M, Bosco D, Bauer C, Morf J, Halban P, Philippe J, Dibner C (2013) Autonomous and self-sustained circadian oscillators displayed in human islet cells. Diabetologia 56:497–507

Sadacca LA, Lamia KA, deLemos AS, Blum B, Weitz CJ (2011) An intrinsic circadian clock of the pancreas is required for normal insulin release and glucose homeostasis in mice. Diabetologia 54:120–124

Stokkan KA, Yamazaki S, Tei H, Sakaki Y, Menaker M (2001) Entrainment of the circadian clock in the liver by feeding. Science 291:490–493

Wang G, Han T, Nijhawan D, Theodoropoulos P, Naidoo J, Yadavalli S, Mirzaei H, Pieper AA, Ready JM, McKnight SL (2014) P7C3 neuroprotective chemicals function by activating the rate-limiting enzyme in NAD salvage. Cell 158:1324–1334

Control of Metabolism by Central and Peripheral Clocks in Drosophila

Amita Sehgal

Abstract Drosophila is a powerful system for the molecular analysis of circadian clocks, providing the first account of how such a clock is generated. It is also proving to be an excellent model to dissect the neural basis of circadian behavior. In addition, clocks are located in peripheral tissues in flies, but much less is known about these clocks and about the physiological processes they control. This chapter describes the use of Drosophila for understanding the circadian control of metabolism. While a clock in the fat body is critical for metabolic function, it is clear that neuronal clocks are also involved. Indeed, synchrony between these clocks is important for reproductive fitness. A complex interplay between circadian and metabolic signals is indicated by the finding that metabolic pathways can even impact rest:activity rhythms controlled by the brain clock. Drosophila may be an optimal system to dissect the nature of these interactions and their importance for organismal fitness and life span.

Genetic analysis of circadian rhythms started with the isolation of the *period (per)* mutants in the fruit fly, *Drosophila melanogaster*, followed by isolation of the *per* gene in the mid 1980s (Bargiello et al. 1984; Jackson et al. 1986; Konopka and Benzer 1971; Reddy et al. 1984; Zehring et al. 1984). Subsequent studies identified the *per* partner, *timeless (tim)*, and the transcriptional feedback loop that we now know lies at the heart of the clock mechanism in all species (Sehgal et al. 1994, 1995). In the Drosophila loop, the Clock (CLK) and cycle (CYC) transcriptional activators promote expression of *per* and *tim* mRNA during the mid to late day but are repressed by feedback activity of PER-TIM in the late night and early morning. Regulated expression and activity of clock proteins in this loop are sustained through post-translational mechanisms, in particular the action of multiple kinases and phosphatases (Zheng and Sehgal 2008, 2012).

Contrary to expectations that clocks would be localized largely, if not exclusively, in the brain, analysis of Drosophila *per* showed that it was expressed in multiple tissues throughout the body (Liu et al. 1988; Saez and Young 1988). Indeed, use of a reporter in which *per* was fused to firefly luciferase showed that

A. Sehgal (✉)
Perelman School of Medicine, University of Pennsylvania, Philadelphia, PA, USA
e-mail: amita@mail.med.upenn.edu

© The Author(s) 2016
P. Sassone-Corsi, Y. Christen (eds.), *A Time for Metabolism and Hormones*,
Research and Perspectives in Endocrine Interactions,
DOI 10.1007/978-3-319-27069-2_4

per was expressed cyclically in most tissues. Analysis of isolated tissues revealed that luciferase activity continued to cycle in the absence of neural connections or systemic signals, indicating the presence of tissue-autonomous clocks (Plautz et al. 1997). Subsequent studies showed that the degree of autonomy varied from tissue to tissue. The Malphigian tubules or fly kidneys, for instance, appeared to be completely autonomous, such that they retained their own "timing" even when transplanted into a host that was synchronized to a different day:night cycle (in other words, a different time zone; Giebultowicz et al. 2000). On the other hand, the clock in the prothoracic gland, which drives a circadian rhythm of eclosion (hatching of adult flies from pupae) in Drosophila, is "slave" to the "master clock in the brain (Myers et al. 2003). Thus, the brain clock is required for eclosion rhythms as well as for maintenance of the prothoracic clock (Myers et al. 2003). In addition, central nervous system signals, in particular the neuropeptide Pigment Dispersing Factor (PDF), modulate the clock in pheromone-producing oenocytes, which regulate mating (Krupp et al. 2013).

The emerging pattern is that of a network of clocks that control many aspects of physiology and depend upon neural function to varying extents. The question is the extent to which Drosophila can be used to study circadian regulation of these different physiological processes and provide an understanding of the circadian system as a whole. This chapter outlines studies directed towards circadian control of metabolism in Drosophila.

Use of Drosophila to Study Behavior and Metabolic Function

As noted above, Drosophila has proved to be an outstanding system to dissect the molecular basis of the clock. Genes first found in Drosophila are now known to be mutated in some human circadian disorders. It is now also clear that Drosophila can be exploited to provide a complete understanding of the neural circuits that drive rhythms in behavior. The *per* and *tim* mutants were isolated through screens that used eclosion behavior as an assay for circadian function. Eclosion is "gated" by the circadian clock to occur around dawn, so while it only occurs once in the life of every fly, it can be monitored as a rhythm in a population. In addition to eclosion, the *per* and *tim* mutants were found to affect rhythms of rest:activity, and subsequently, in particular with the development of high throughput systems for monitoring locomotor activity, the field shifted to almost exclusively using rest:activity as a readout of internal clock function. Through work done in several laboratories, we now have a fairly good understanding of the clock neurons in the brain that drive rhythms of rest:activity (Nitabach and Taghert 2008). Interestingly, different subsets of neurons are required for different aspects of the overt rhythm, for instance. for the morning and evening peaks of locomotor activity. In addition, we recently identified a neural circuit that connects the clock neurons to other brain cells required for rhythmic rest:activity (Cavanaugh et al. 2014). It seems likely that, in the near future, we will be able to trace the passage of time-of-day signals all the way from the clock to the motor neurons that drive activity.

Until ~2008, little to no work had been done on circadian metabolism in Drosophila. However, flies have been used for general studies of metabolism, and are particularly useful as a model for aging, which is influenced strongly by metabolic parameters (Katewa and Kapahi 2010). As circadian regulation may be relevant for aging, we undertook to address links between metabolism and the circadian system.

The Drosophila Fat Body Contains a Clock that Regulates a Rhythm of Feeding

As we were accustomed to monitoring behavior in Drosophila, our studies of metabolic function also started with measurements of a metabolism-influenced behavior. We assayed food intake at different times of day and found that flies display a circadian rhythm of feeding such that food intake occurs maximally in the morning hours (Xu et al. 2008). A later study identified an additional peak of feeding that occurs later in the day and confirmed that nighttime hours of quiescence are associated with reduced food intake. As required of an endogenously driven rhythm, the rhythm of feeding persists in the dark, i.e., in the absence of environmental cycles. Also, it is eliminated in the dark in flies lacking the *Clk* gene, demonstrating that it is under the control of the molecular clock mechanism described above (Xu et al. 2008).

To address the regulation of the feeding rhythm, we considered a role for the fat body, as this is a major metabolic tissue in Drosophila and is generally considered the functional equivalent of the liver. We found that clock genes, specifically *tim*, were expressed in the fat body and displayed a daily rhythm (Xu et al. 2008). To determine if this cycling was driven by a clock in the fat body, as opposed to signals from elsewhere, we disrupted the fat body clock by transgenically expressing a dominant negative version of the CLK protein. This manipulation abolished *tim* cycling, indicating that it depends upon a clock in the fat body. Interestingly, disruption of the fat body clock also affected the phase of the feeding rhythm, such that flies now showed maximal food consumption in the evening hours (Xu et al. 2008). The fact that the feeding rhythm was not abolished suggests that clocks in other tissues can also drive this rhythm.

Fat Body and Neuronal Clocks Coordinately Regulate Metabolic Parameters

We found that loss of the fat body clock did not just affect the feeding rhythm but also overall food intake (Xu et al. 2008). Food consumption was higher at all times of day relative to controls. Reasoning that increased food consumption increases sources of energy and therefore might be protective in adverse conditions of low nutrient availability, we tested flies lacking a fat body clock in starvation assays. To our surprise, we found that they were actually more sensitive to starvation and so died earlier than their wild type counterparts. This finding suggested that the increased food consumption was not increasing nutrient stores but was perhaps occurring in response to low endogenous levels of nutrients. Indeed, we found that glycogen and triglyceride levels were low in flies that lacked a clock in the fat body.

These results were unexpected because clock mutants, in other words flies lacking clocks in all tissues, do not show obvious metabolic phenotypes. The defects seen when only the fat body clock was ablated suggested that clocks in other tissues might have opposing effects on metabolic parameters. Neurons appeared to be good candidates for housing such clocks, as the brain is known to regulate metabolic activity, and so we disrupted clock function in neurons. We used the same tool as for the fat body clock (dominant negative clock proteins) and confirmed that neuronal clocks were disrupted by monitoring rest:activity behavior. As expected, rest:activity was arrhythmic. Measurement of metabolic parameters showed that nutrient stores, triglycerides and glycogen, were higher in flies with disrupted neuronal clocks than in wild type controls (Xu et al. 2008). As might be predicted, loss of neuronal clocks also increased resistance to starvation.

These data indicate that the fat body and the neuronal clock oppose each other in the control of metabolic function (Fig. 1). Typically, the fat body clock suppresses feeding, promotes storage of nutrients and increases resistance to starvation. Thus, loss of the fat body clock results in increased feeding, lower nutrient stores and sensitivity to starvation. Conversely, neurons are very metabolically active, and so clocks in these promote feeding, depletion of energy stores and sensitivity to starvation. All these functions are likely reversed when neuronal clocks are lost.

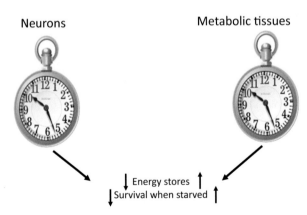

Fig. 1 Neuronal and metabolic clocks have opposing effects on metabolic parameters. These effects are predicted based upon phenotypes obtained by disrupting neuronal or fat body (metabolic) clocks. Disruption of neuronal clocks increases glycogen and trigyceride stores and promotes survival in response to starvation, whereas disruption of the fat body clock decreases glycogen and trigyceride stores, increases feeding and decreases survival upon starvation

We documented increased nutrients and resistance to starvation in the absence of neuronal clocks but were unable to reliably quantify food intake, as this was so low.

In a subsequent study, we identified a specific group of neuron that regulate triglyceride levels (DiAngelo et al. 2011). These are the central clock neurons in the brain, which are critical for rest:activity rhythms. Interestingly, though, the effects of these neurons on triglyceride levels are separable from their effects on rest: activity.

Rhythmic Gene Expression in the Fat Body Is Controlled Largely, but not Exclusively, by the Fat Body Clock

To address the mechanisms by which the fat body clock regulates metabolic homeostasis, we sought to identify the genes expressed rhythmically in this tissue (Fig. 2). To this end, we collected tissue every 2 h around the clock over a 2-day period and profiled gene expression using microarrays (Xu et al. 2011). Simultaneously, we collected samples every 4 h from flies lacking a fat body clock due to expression of a dominant negative form of the CLK protein. We found that expression of many genes is cyclic in the fat body. Interestingly, several of these continue to cycle when the fat body clock is ablated, suggesting the influence of other factors, either the light:dark cycle or clocks elsewhere. In recent work, we have found that clocks in other tissues are required for at least some of the rhythmic cycling in the fat body.

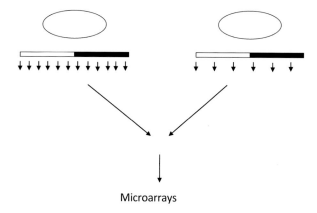

Fig. 2 Circadian gene expression in the fat body: The protocol shown was followed to assay circadian gene expression in the fat body. Fat bodies were collected at 2-h intervals over a 48-h cycle in wild type flies and at 4-h intervals in flies lacking a fat body clock. Several classes of genes were found to cycle

Microarrays

A Restricted Feeding Paradigm Resets the Phase of Cyclic Gene Expression in the Fat Body but not in the Brain

The genes expressed cyclically in the fat body fall into many different functional categories, including lipid synthesis (in particular, fatty acid elongation), lipid breakdown, steroid hormone metabolism and immune function. The peak of gene expression for these different processes tended to occur at different times of day. To determine if temporal separation of gene expression by the clock was important for metabolic physiology, we sought to disrupt this temporal relationship. Reasoning that the time of feeding might be important for the peak in metabolic gene expression, but perhaps not for expression of immune genes, we restricted food to a time of day when feeding was typically less (6 h in the early evening) and we examined circadian gene expression (Xu et al. 2011). We found that the time of feeding was indeed important, in fact even more than predicted. Thus, the clock in the fat body was reset by the time of feeding, which led to a reset of all downstream cycling genes.

Restricted feeding (RF) only changed the phase of gene expression if it occurred at the wrong time of day. If food was restricted to a time that corresponded to the normal daily peak of feeding, then the phase was maintained and the amplitude of the rhythm became stronger (note that normally the amplitude is low in constant darkness). On the other hand, RF had no effect on circadian gene expression in the brain (Xu et al. 2011).

Decoupling Peripheral and Brain Tissues Decreases Reproductive Fitness

As discussed above, a RF paradigm desynchronizes brain and fat body clocks as it resets the fat body, but not the brain clock. To determine if this process had physiological consequences, we monitored egg laying as a measure of reproductive fitness in animals maintained on RF. To exclude any influence of the duration of feeding, we compared egg production by flies fed for 6 h daily at the time they would normally eat with those fed for 6 h at the wrong time of day (Xu et al. 2011). Measurements of food intake showed equal food consumption in both groups, indicating that 18 h of starvation promoted equivalent feeding regardless of circadian time.

We found that flies fed at the wrong time laid fewer eggs than those fed at the correct time. However, these differences were not noted in a *Clk* mutant, indicating that they reflected an interaction of the time of feeding with endogenous clocks (Xu et al. 2011). We surmise that desynchrony of brain and peripheral clocks, achieved by an RF paradigm, reduced reproductive success.

Metabolic Signals Also Affect Clocks in the Brain

While this chapter focuses on the circadian control of metabolism, we have also uncovered effects of metabolic signals of central clock function and rest:activity behavior. We found that the FOXO protein, a well-known component of metabolic pathways, is expressed in the fat body but can influence the brain clock's response to oxidative stress (Zheng et al. 2007). We also found that manipulations of the TOR-Akt pathway alter periodicity of rest:activity rhythms in parallel with effects on the molecular clock in brain neurons (Zheng and Sehgal 2010). Thus, metabolism and circadian clock interact on multiple levels, with consequences in both directions.

References

Bargiello TA, Jackson FR, Young MW (1984) Restoration of circadian behavioural rhythms by gene transfer in Drosophila. Nature 312:752–754

Cavanaugh DJ, Geratowski JD, Wooltorton JR, Spaethling JM, Hector CE, Zheng X, Johnson EC, Eberwine JH, Sehgal A (2014) Identification of a circadian output circuit for rest:activity rhythms in Drosophila. Cell 157:689–701

DiAngelo JR, Erion R, Crocker A, Sehgal A (2011) The central clock neurons regulate lipid storage in Drosophila. PLoS One 6:e19921

Giebultowicz JM, Stanewsky R, Hall JC, Hege DM (2000) Transplanted Drosophila excretory tubules maintain circadian clock cycling out of phase with the host. Curr Biol 10:107–110

Jackson FR, Bargiello TA, Yun SH, Young MW (1986) Product of per locus of Drosophila shares homology with proteoglycans. Nature 320:185–188

Katewa SD, Kapahi P (2010) Dietary restriction and aging, 2009. Aging Cell 9:105–112

Konopka RJ, Benzer S (1971) Clock mutants of *Drosophila melanogaster*. Proc Natl Acad Sci USA 68:2112–2116

Krupp JJ, Billeter JC, Wong A, Choi C, Nitabach MN, Levine JD (2013) Pigment-dispersing factor modulates pheromone production in clock cells that influence mating in drosophila. Neuron 79:54–68

Liu X, Lorenz L, Yu QN, Hall JC, Rosbash M (1988) Spatial and temporal expression of the period gene in Drosophila melanogaster. Genes Dev 2:228–238

Myers EM, Yu J, Sehgal A (2003) Circadian control of eclosion. Interaction between a central and peripheral clock in Drosophila melanogaster. Curr Biol 13:526–533

Nitabach MN, Taghert PH (2008) Organization of the Drosophila circadian control circuit. Curr Biol 18:R84–R93

Plautz JD, Kaneko M, Hall JC, Kay SA (1997) Independent photoreceptive circadian clocks throughout Drosophila. Science 278:1632–1635

Reddy P, Zehring WA, Wheeler DA, Pirrotta V, Hadfield C et al (1984) Molecular analysis of the period locus in Drosophila melanogaster and identification of a transcript involved in biological rhythms. Cell 38:701–710

Saez L, Young MW (1988) In situ localization of the per clock protein during development of Drosophila melanogaster. Mol Cell Biol 8:5378–5385

Sehgal A, Price JL, Man B, Young MW (1994) Loss of circadian behavioral rhythms and per RNA oscillations in the Drosophila mutant timeless. Science 263:1603–1606

Sehgal A, Rothenfluh-Hilfiker A, Hunter-Ensor M, Chen Y, Myers MP, Young MW (1995) Rhythmic expression of timeless: a basis for promoting circadian cycles in period gene auto-regulation. Science 270:808–810

Xu K, Zheng X, Sehgal A (2008) Regulation of feeding and metabolism by neuronal and peripheral clocks in Drosophila. Cell Metab 8:289–300

Xu K, DiAngelo JR, Hughes ME, Hogenesch JB, Sehgal A (2011) The circadian clock interacts with metabolic physiology to influence reproductive fitness. Cell Metab 13:639–654

Zehring WA, Wheeler DA, Reddy P, Konopka RJ, Kyriacou CP et al (1984) P-element transformation with period locus DNA restores rhythmicity to mutant, arrhythmic Drosophila melanogaster. Cell 39:369–376

Zheng X, Sehgal A (2008) Probing the relative importance of molecular oscillations in the circadian clock. Genetics 178:1147–1155

Zheng X, Sehgal A (2010) AKT and TOR signaling set the pace of the circadian pacemaker. Curr Biol 20:1203–1208

Zheng X, Sehgal A (2012) Speed control: cogs and gears that drive the circadian clock. Trends Neurosci 35:574–585

Zheng X, Yang Z, Yue Z, Alvarez JD, Sehgal A (2007) FOXO and insulin signaling regulate sensitivity of the circadian clock to oxidative stress. Proc Natl Acad Sci USA 104:15899–15904

Circadian Post-transcriptional Control of Metabolism

Carla B. Green

Abstract Circadian clocks control thousands of genes, which ultimately generate rhythms in signaling pathways, metabolism, tissue physiology and behavior. Although rhythmic transcription plays a critical role in generating these rhythmic gene expression patterns, recent evidence has shown that post-transcriptional mechanisms are also important. Here we describe studies showing that regulation of mRNA poly(A) tail length is under circadian control and that these changes contribute to rhythmic protein expression independently of transcription. Nocturnin, a circadian deadenylase that shortens poly(A) tails, contributes to this type of circadian post-transcriptional regulation. The importance of tail-shortening by Nocturnin is evident from the phenotype of mice lacking Nocturnin, which exhibit resistance to diet-induced obesity and other metabolic changes.

Introduction

Circadian clocks regulate and coordinate rhythms in behavior, physiology, bio-chemistry and gene expression in mammals (Pittendrigh 1981a, b; Akhtar et al. 2002; Panda et al. 2002; Storch et al. 2002; Ueda et al. 2002; Duffield 2003; Welsh et al. 2004; Reddy et al. 2006), allowing animals to synchronize appropriately to the environmental light:dark cycles. The mammalian circadian clock is composed of an intracellular feedback mechanism in which interlocking transcriptional-translational feedback loops generate the 24-h rhythms (reviewed in Lowrey and Takahashi 2004; Takahashi et al. 2008) and drive rhythms of 5–10 % of genes in a cell type-specific manner (Duffield 2003; Rey et al. 2011; Koike et al. 2012; Menet et al. 2012). This extensive control over mRNA expression results in rhythmicity of many cellular pathways, including many aspects of metabolism. Mutations that alter the clock have broad negative effects on the

C.B. Green (✉)
Department of Neuroscience, University of Texas Southwestern Medical Center, Dallas, TX, USA
e-mail: Carla.Green@utsouthwestern.edu

© The Author(s) 2016
P. Sassone-Corsi, Y. Christen (eds.), *A Time for Metabolism and Hormones*,
Research and Perspectives in Endocrine Interactions,
DOI 10.1007/978-3-319-27069-2_5

organism, including insulin resistance and obesity (Rudic et al. 2004; Shimba et al. 2005; Turek et al. 2005; Green et al. 2008), some types of cancer (Fu et al. 2002; Gorbacheva et al. 2005; Hoffman et al. 2009, 2010a, b; Ozturk et al. 2009; Kang et al. 2010), cardiovascular disease (Curtis et al. 2007; Reilly et al. 2007) and sleep and affective disorders (Toh et al. 2001; Kripke et al. 2009; Srinivasan et al. 2009; Touma et al. 2009). Therefore, an understanding of the molecular mechanism of clocks in mammals is critical for the understanding and treatment of human health.

The components of the central circadian clock are transcriptional activators and repressors, and cyclic activation and repression drive the oscillation that comprises the pacemaker and generates the 24-h periodicity. In addition, these proteins drive rhythms in many other genes, through both direct and indirect transcriptional mechanisms. Although this transcriptional control is a major contributor to the resulting rhythms in mRNA levels, a number of recent studies have demonstrated that post-transcriptional regulation also must play an important role. For example, a large percent of rhythmic mRNAs in liver do not have rhythmic pre-RNAs (Koike et al. 2012; Menet et al. 2012) and, in mouse liver, almost 50 % of the rhythmic proteins do not have rhythmic steady-state mRNA levels (Reddy et al. 2006). Moreover, circadian rhythms can exist in red blood cells devoid of nuclei (O'Neill and Reddy 2011; O'Neill et al. 2011). Therefore, regulatory mechanisms beyond transcription can also drive rhythmic physiology.

Post-transcriptional Mechanisms

Although transcription drives mRNA synthesis, the ultimate protein expression patterns also reflect regulation at many other levels (Fig. 1). Even as the mRNA is being transcribed, large complexes of proteins associate with the nascent transcript and regulate the efficiency and pattern of splicing, the choice of 3′-end cleavage site and polyadenylation (Pawlicki and Steitz 2010). The mature transcript undergoes further regulation during nuclear export, cytoplasmic localization, RNA stability and translation. The importance of post-transcriptional regulation has become clear over the last decade, with the discovery of many RNA binding proteins, specific types of ribonucleases, and the extensive machinery that conducts microRNA-mediated control of mRNA stability and translation. Although significant progress has been made in this area, understanding of post-transcriptional mechanisms still lags behind that of transcriptional processes.

The poly(A) tails at the 3′-end of most eukaryotic mRNAs are thought to be important for controlling translatability and stability, and one post-transcriptional regulatory mechanism is to modulate the length of these tails. Indeed, regulation of poly(A) tail length has been shown to play critical roles in many biological processes, including oocyte maturation, mitotic cell cycle progression, cellular senescence and synaptic plasticity (Gebauer et al. 1994; Groisman et al. 2002, 2006; Huang et al. 2002, 2006; Novoa et al. 2010). Changes in poly(A) tail length can occur at many points during the lifetime of an mRNA. Long poly(A) tails of

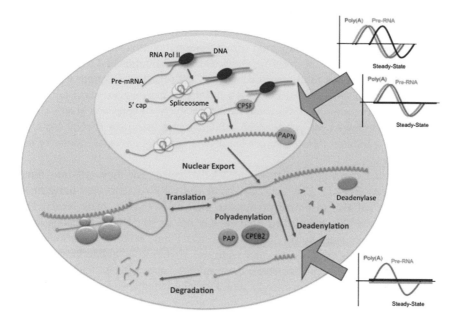

Fig. 1 Post-transcriptional regulation controls expression at many steps throughout the lifetime of the mRNA, and rhythms in poly(A) tail length can result from transcriptionally coupled mechanisms and cytoplasmic mechanisms (Kojima et al. 2012)

about 250 nt are initially added to the nascent transcript in the nucleus following the 3′-end cleavage (Kuhn and Wahle 2004). Following export out of the nucleus, a protein called cytoplasmic poly(A)-binding protein (PABPC) binds to the tail and stabilizes the mRNA. Through direct interactions with the translation-initiation factor eIF4G, which in turn binds to the cap-binding protein eIF4E, PABPC is thought to facilitate translation initiation by forming a "closed-loop" circular structure (Kuhn and Wahle 2004). Removal or shortening of the tail by a specific class of ribonucleases called deadenylases can, in turn, result in translational silencing and, in some cases, mRNA degradation. Alternatively, cytoplasmic polyadenylation can, in some cases, lengthen the tail of an mRNA that was previously shortened and stabilize it and render it translationally competent (Richter 2007).

Circadian Control of poly(A) Tail Length

Daily variations in poly(A) tail length were reported for two mRNAs (Robinson et al. 1988; Gerstner et al. 2012), causing us to wonder whether the circadian clock uses this mechanism more broadly to regulate gene expression post-transcriptionally. Therefore, we fractionated mRNAs from mouse livers collected

at various circadian times into pools of mRNAs with short (~50 nt) and long
(>100 nt) poly(A) tails (Kojima et al. 2012) using a modification of an oligo
(dT) affinity chromatography method with differential elution stringencies (Meijer
et al. 2007). These pools of mRNAs, along with a non-fractionated total poly(A)+
control, were subjected to microarray analysis, and relative tail-length was determined by the ratio of expression of each mRNA in the long vs. short tail pools
(normalized for expression level using the total poly(A)+ expression level). Using
this method, followed by independent validation, we identified several hundred
mRNAs that exhibited robust changes in poly(A) tail length over the course of the
circadian day.

Further characterization of these mRNAs revealed that they fell into three
general classes (Fig. 1). The first class contained mRNAs that were transcribed
rhythmically and also exhibited rhythms in their overall steady-state levels. The
second class was also transcribed rhythmically but these mRNAs had long half-
lives and therefore were not rhythmic at the steady-state level. The third class of
mRNAs with rhythmic poly(A) tails were not rhythmic at either the level of
synthesis or at the steady-state level; these mRNAs appear to be long-lived and
have poly(A) tails that are cyclically shortened and lengthened in the cytoplasm.
Strikingly, in all the cases we tested, the poly(A) tail rhythms correlated with
rhythmic protein levels, even in the cases where the steady-state levels of the
mRNAs were not changing. These data suggest that circadian changes in poly
(A) tail length can significantly contribute to rhythmic protein synthesis, independent of transcription.

Nocturnin Is a Circadian Deadenylase

The mechanism by which the clock controls poly(A) tail length is not well understood and appears to involve different mechanisms at different circadian phases
(Kojima et al. 2012). However, one strong candidate is the deadenylase Nocturnin
(gene name, *Ccrn4l*; Green and Besharse 1996; Baggs and Green 2003), which is
robustly rhythmic in many mouse tissues, with peak expression in the middle of the
night (Wang et al. 2001; Garbarino-Pico et al. 2007; Kojima et al. 2010). Nocturnin
is a member of the superfamily of deadenylases that includes CCR4, Nocturnin,
Angel, and 2′PDE (Goldstrohm and Wickens 2008; Godwin et al. 2013), but
Nocturnin has a distinct amino-terminus from the other members. Nocturnin is
also unique among all deadenylases in its characteristic high amplitude rhythms,
with nighttime peaks (most of the other deadenylases are arrhythmic or have very
low amplitude rhythms that peak in the day) (Kojima et al. 2012). In addition,
Nocturnin is unique in that it is an immediate early gene that is acutely induced by
many stimuli (Garbarino-Pico et al. 2007). Given the difference in temporal and
spatial expression patterns of the deadenylases (Yamashita et al. 2005; Morita
et al. 2007; Wagner et al. 2007; Kojima et al. 2012) and the different phenotypes
caused by disrupting specific deadenylases (Molin and Puisieux 2005; Morris

et al. 2005; Green et al. 2007; Morita et al. 2007; Washio-Oikawa et al. 2007), it is likely that each deadenylase targets a specific set of transcripts, although the identities of these transcripts and the mechanisms by which they are targeted by a particular deadenylase are not well characterized.

Loss of Nocturnin Results in Broad Metabolic Changes

The importance of Nocturnin's contribution to circadian changes in poly(A) tail length was tested by generating mice lacking Nocturnin (*Noc−/−*; Green et al. 2007). These mice appeared normal and healthy when raised in standard conditions and bred well. However, when raised on a Western-style high fat diet, the *Noc−/−* mice did not gain weight at the same rate as the wild-type mice and remained lean whereas the wild-type mice became obese (Fig. 2). The *Noc−/−* mice had smaller fat pads and were protected from hepatic steatosis. Despite this resistance to diet-induced obesity, the *Noc−/−* mice did not eat less, were not more active, and did not show significant changes in whole body respiration as measured in metabolic cages (Green et al. 2007; Douris and Green 2008). These mice did, however, have changes in mRNA expression levels of many key metabolic regulators in the liver, often showing loss of rhythmicity of normally rhythmic genes. Nocturnin is likely not part of the core circadian mechanism, because the *Noc−/−* mice had normal circadian locomotor rhythms and normal expression of the core clock genes in the liver. However, it is directly regulated by the core circadian transcription factor heterodimer CLOCK/BMAL1 and is, therefore, a direct output of the intracellular core circadian loop. In addition, it is regulated by systemic circadian signals, likely originating directly or indirectly from the core circadian pacemaker in the suprachiasmatic nucleus in the hypothalamus, because Nocturnin is one of only a few dozen rhythmic genes that maintain rhythmicity following

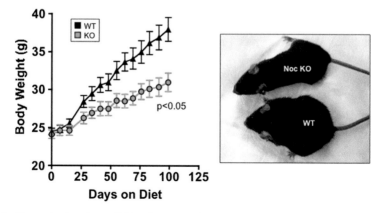

Fig. 2 Nocturnin knockout (KO) mice are resistant to diet-induced obesity. *WT* wild-type (Modified from Green et al. 2007)

genetic disruption of the clock, specifically in the liver of mice (Kornmann et al. 2007).

Some clues to the mechanism behind the lean phenotype observed in the *Noc*−/− mice came from examination of Nocturnin's role in the small intestine (Douris et al. 2011). Nocturnin is expressed throughout the digestive tract, but with particularly high levels in the upper part of the small intestine. As in other tissues, it is robustly rhythmic, peaking during the night—the time of maximal food intake in the nocturnal mouse. Pan and Hussain (2009) had previously shown that many of the transporters involved in macronutrient absorption by the intestinal enterocytes were under the control of the circadian clock. Accordingly, we found that lipid absorption in the wild-type mice was strongly circadian, with rapid appearance of newly ingested lipoprotein particles into the circulation when the mice were gavaged with olive oil at night, but slow and limited appearance when olive oil was administered during the day. In contrast, the *Noc*−/− mice had no rhythm in absorption and exhibited slow "daytime-like" absorption profiles following gavage given both night and day (Douris et al. 2011; Stubblefield et al. 2012). Furthermore, the enterocytes in the *Noc*−/− mice accumulated large cytoplasmic lipid droplets, suggesting that dietary lipids were stored in these cells when Nocturnin was not present—at all times in the *Noc*−/− cells or during the daytime in wild-type mice. The mechanism by which this deadenylase regulates dietary lipid absorption is not clear, but several mRNAs that encode proteins involved in lipid droplet formation, breakdown and chylomicron assembly are dysregulated in the *Noc*−/− intestine, and some of these may be direct targets of Nocturnin deadenylase activity.

Nocturnin also plays important roles in other metabolically relevant tissues. In bone, Nocturnin interacts with a specific long isoform of *Igf1* mRNA, suppressing its expression (Kawai et al. 2010a). In bone-marrow stromal cells, Nocturnin is acutely induced more than 30-fold by the peroxisome proliferator-activated receptor gamma (PPAR-gamma) agonist rosiglitazone, and *Noc*−/− mice have reduced marrow adiposity and high bone mass (Kawai et al. 2010b). In addition, overexpression of Nocturnin enhances adipogeneis in preadipocyte 3T3-L1 cells and negatively regulates osteogenesis in mouse osteoblastic MC3T3-E1 cells (Kawai et al. 2010b). Together these data suggest that Nocturnin plays an important role in the mesenchymal stem-cell lineage allocation that may ultimately influence adipogenesis and body composition.

Conclusions

The large contribution of post-transcriptional regulation to the generation and modulation of rhythmic mRNA and protein profiles has recently become apparent largely thanks to the use of genome-wide interrogation of rhythmic mRNA expression and transcriptional and post-transcriptional states. The ongoing development of innovative high-throughput methods for analyzing various nuances of gene expression (TAIL-seq, GRO-seq, CLIP-seq, and many more) will undoubtedly yield ever more information about how the clock controls the many layers of gene expression that drive the complex rhythmic physiology and behavior of mammals. We expect that new modes of post-transcriptional regulatory mechanisms will be uncovered and that these will be shown to play an important role in shaping these rhythms.

Nocturnin is likely only one of many post-transcriptional modulators that contribute to circadian expression profiles, but the profound metabolic phenotype in the *Noc*−/− mice shows that it is playing an important role in regulating circadian metabolic profiles. However, to understand how loss of Nocturnin causes these phenotypes, it will be critical to identify the relevant Nocturnin target mRNAs and to uncover how these target mRNAs are recognized by Nocturnin. Finally, the exact function of deadenylation by Nocturnin is still not clear. Although removal of tails has long been thought to target them for decay, it has recently been discovered that many mRNAs are maintained in the cell in short-tailed states that are quite stable. Are these short-tailed mRNAs translationally silent and waiting to have their tails lengthened in response to the appropriate signal or do they have some other function? Only time will tell...

References

Akhtar RA, Reddy AB, Maywood ES, Clayton JD, King VM, Smith AG, Gant TW, Hastings MH, Kyriacou CP (2002) Circadian cycling of the mouse liver transcriptome, as revealed by cDNA microarray, is driven by the suprachiasmatic nucleus. Curr Biol 12:540–550

Baggs JE, Green CB (2003) Nocturnin, a deadenylase in Xenopus laevis retina: a mechanism for posttranscriptional control of circadian-related mRNA. Curr Biol 13:189–198

Curtis AM, Cheng Y, Kapoor S, Reilly D, Price TS, Fitzgerald GA (2007) Circadian variation of blood pressure and the vascular response to asynchronous stress. Proc Natl Acad Sci USA 104:3450–3455

Douris N, Green CB (2008) NOC out the fat: a short review of the circadian deadenylase nocturnin. Ann Med 40:622–626

Douris N, Kojima S, Pan X, Lerch-Gaggl AF, Duong SQ, Hussain MM, Green CB (2011) Nocturnin regulates circadian trafficking of dietary lipid in intestinal enterocytes. Curr Biol 21:1347–1355

Duffield GE (2003) DNA microarray analyses of circadian timing: the genomic basis of biological time. J Neuroendocrinol 15:991–1002

Fu L, Pelicano H, Liu J, Huang P, Lee C (2002) The circadian gene Period2 plays an important role in tumor suppression and DNA damage response in vivo. Cell 111:41–50

Garbarino-Pico E, Niu S, Rollag MD, Strayer CA, Besharse JC, Green CB (2007) Immediate early response of the circadian polyA ribonuclease nocturnin to two extracellular stimuli. RNA 13:745–755

Gebauer F, Xu W, Cooper GM, Richter JD (1994) Translational control by cytoplasmic polyadenylation of c-mos mRNA is necessary for oocyte maturation in the mouse. EMBO J 13:5712–5720

Gerstner JR, Vanderheyden WM, LaVaute T, Westmark CJ, Rouhana L, Pack AI, Wickens M, Landry CF (2012) Time of day regulates subcellular trafficking, tripartite synaptic localization, and polyadenylation of the astrocytic Fabp7 mRNA. J Neurosci 32:1383–1394

Godwin AR, Kojima S, Green CB, Wilusz J (2013) Kiss your tail goodbye: the role of PARN, nocturnin, and angel deadenylases in mRNA biology. Biochim Biophys Acta 1829:571–579

Goldstrohm AC, Wickens M (2008) Multifunctional deadenylase complexes diversify mRNA control. Nat Rev Mol Cell Biol 9:337–344

Gorbacheva VY, Kondratov RV, Zhang R, Cherukuri S, Gudkov AV, Takahashi JS, Antoch MP (2005) Circadian sensitivity to the chemotherapeutic agent cyclophosphamide depends on the functional status of the CLOCK/BMAL1 transactivation complex. Proc Natl Acad Sci USA 102:3407–3412

Green CB, Besharse JB (1996) Identification of a novel vertebrate circadian clock-regulated gene encoding the protein nocturnin. Proc Natl Acad Sci USA 93:14884–14888

Green CB, Douris N, Kojima S, Strayer CA, Fogerty J, Lourim D, Keller SR, Besharse JC (2007) Loss of nocturnin, a circadian deadenylase, confers resistance to hepatic steatosis and diet-induced obesity. Proc Natl Acad Sci USA 104:9888–9893

Green CB, Takahashi JS, Bass J (2008) The meter of metabolism. Cell 134:728–742

Groisman I, Jung MY, Sarkissian M, Cao Q, Richter JD (2002) Translational control of the embryonic cell cycle. Cell 109:473–483

Groisman I, Ivshina M, Marin V, Kennedy NJ, Davis RJ, Richter JD (2006) Control of cellular senescence by CPEB. Genes Dev 20:2701–2712

Hoffman AE, Zheng T, Stevens RG, Ba Y, Zhang Y, Leaderer D, Yi C, Holford TR, Zhu Y (2009) Clock-cancer connection in non-Hodgkin's lymphoma: a genetic association study and pathway analysis of the circadian gene cryptochrome 2. Cancer Res 69:3605–3613

Hoffman AE, Zheng T, Ba Y, Stevens RG, Yi CH, Leaderer D, Zhu Y (2010a) Phenotypic effects of the circadian gene cryptochrome 2 on cancer-related pathways. BMC Cancer 10:110

Hoffman AE, Zheng T, Yi CH, Stevens RG, Ba Y, Zhang Y, Leaderer D, Holford T, Hansen J, Zhu Y (2010b) The core circadian gene cryptochrome 2 influences breast cancer risk, possibly by mediating hormone signaling. Cancer Prev Res (Phila) 3:539–548

Huang YS, Jung MY, Sarkissian M, Richter JD (2002) N-methyl-D-aspartate receptor signaling results in Aurora kinase-catalyzed CPEB phosphorylation and alpha CaMKII mRNA polyadenylation at synapses. EMBO J 21:2139–2148

Huang YS, Kan MC, Lin CL, Richter JD (2006) CPEB3 and CPEB4 in neurons: analysis of RNA-binding specificity and translational control of AMPA receptor GluR2 mRNA. EMBO J 25:4865–4876

Kang TH, Lindsey-Boltz LA, Reardon JT, Sancar A (2010) Circadian control of XPA and excision repair of cisplatin-DNA damage by cryptochrome and HERC2 ubiquitin ligase. Proc Natl Acad Sci USA 107:4890–4895

Kawai M, Delany AM, Green CB, Adamo ML, Rosen CJ (2010a) Nocturnin suppresses igf1 expression in bone by targeting the 3′ untranslated region of igf1 mRNA. Endocrinology 151:4861–4870

Kawai M, Green CB, Lecka-Czernik B, Douris N, Gilbert MR, Kojima S, Ackert-Bicknell C, Garg N, Horowitz MC, Adamo ML, Clemmons DR, Rosen CJ (2010b) A circadian-regulated gene, nocturnin, promotes adipogenesis by stimulating PPAR-gamma nuclear translocation. Proc Natl Acad Sci USA 107:10508–10513

Koike N, Yoo SH, Huang HC, Kumar V, Lee C, Kim TK, Takahashi JS (2012) Transcriptional architecture and chromatin landscape of the core circadian clock in mammals. Science 338:349–354

Kojima S, Gatfield D, Esau CC, Green CB (2010) MicroRNA-122 modulates the rhythmic expression profile of the circadian deadenylase nocturnin in mouse liver. PLoS One 5:e11264

Kojima S, Sher-Chen EL, Green CB (2012) Circadian control of mRNA polyadenylation dynamics regulates rhythmic protein expression. Genes Dev 26:2724–2736

Kornmann B, Schaad O, Bujard H, Takahashi JS, Schibler U (2007) System-driven and oscillator-dependent circadian transcription in mice with a conditionally active liver clock. PLoS Biol 5: e34

Kripke DF, Nievergelt CM, Joo E, Shekhtman T, Kelsoe JR (2009) Circadian polymorphisms associated with affective disorders. J Circadian Rhythms 7:2

Kuhn U, Wahle E (2004) Structure and function of poly(A) binding proteins. Biochim Biophys Acta 1678:67–84

Lowrey PL, Takahashi JS (2004) Mammalian circadian biology: elucidating genome-wide levels of temporal organization. Annu Rev Genomics Hum Genet 5:407–441

Meijer HA, Bushell M, Hill K, Gant TW, Willis AE, Jones P, de Moor CH (2007) A novel method for poly(A) fractionation reveals a large population of mRNAs with a short poly(A) tail in mammalian cells. Nucleic Acids Res 35:e132

Menet JS, Rodriguez J, Abruzzi KC, Rosbash M (2012) Nascent-seq reveals novel features of mouse circadian transcriptional regulation. eLife 1:e00011

Molin L, Puisieux A (2005) C. elegans homologue of the Caf1 gene, which encodes a subunit of the CCR4-NOT complex, is essential for embryonic and larval development and for meiotic progression. Gene 358:73–81

Morita M, Suzuki T, Nakamura T, Yokoyama K, Miyasaka T, Yamamoto T (2007) Depletion of mammalian CCR4b deadenylase triggers elevation of the p27Kip1 mRNA level and impairs cell growth. Mol Cell Biol 27:4980–4990

Morris JZ, Hong A, Lilly MA, Lehmann R (2005) twin, a CCR4 homolog, regulates cyclin poly (A) tail length to permit *Drosophila oogenesis*. Development 132:1165–1174

Novoa I, Gallego J, Ferreira PG, Mendez R (2010) Mitotic cell-cycle progression is regulated by CPEB1 and CPEB4-dependent translational control. Nat Cell Biol 12:447–456

O'Neill JS, Reddy AB (2011) Circadian clocks in human red blood cells. Nature 469:498–503

O'Neill JS, van Ooijen G, Dixon LE, Troein C, Corellou F, Bouget FY, Reddy AB, Millar AJ (2011) Circadian rhythms persist without transcription in a eukaryote. Nature 469:554–558

Ozturk N, Lee JH, Gaddameedhi S, Sancar A (2009) Loss of cryptochrome reduces cancer risk in p53 mutant mice. Proc Natl Acad Sci USA 106:2841–2846

Pan X, Hussain MM (2009) Clock is important for food and circadian regulation of macronutrient absorption in mice. J Lipid Res 50:1800–1813

Panda S, Antoch MP, Miller BH, Su AI, Schook AB, Straume M, Schultz PG, Kay SA, Takahashi JS, Hogenesch JB (2002) Coordinated transcription of key pathways in the mouse by the circadian clock. Cell 109:307–320

Pawlicki JM, Steitz JA (2010) Nuclear networking fashions pre-messenger RNA and primary microRNA transcripts for function. Trends Cell Biol 20:52–61

Pittendrigh CS (1981a) Circadian systems: general perspective. In: Aschoff J (ed) Biological rhythms, vol 4, 1st edn, Handbook of behavioral neurobiology. Plenum, New York, pp 57–80

Pittendrigh CS (1981b) Circadian systems: entrainment. In: Ascholl J (ed) Biological rhythms, vol 4, 1st edn, Handbook of behavioral neurobiology. Plenum, New York, pp 95–124

Reddy AB, Karp NA, Maywood ES, Sage EA, Deery M, O'Neill JS, Wong GK, Chesham J, Odell M, Lilley KS, Kyriacou CP, Hastings MH (2006) Circadian orchestration of the hepatic proteome. Curr Biol 16:1107–1115

Reilly DF, Westgate EJ, FitzGerald GA (2007) Peripheral circadian clocks in the vasculature. Arterioscler Thromb Vasc Biol 27:1694–1705

Rey G, Cesbron F, Rougemont J, Reinke H, Brunner M, Naef F (2011) Genome-wide and phase-specific DNA-binding rhythms of BMAL1 control circadian output functions in mouse liver. PLoS Biol 9:e1000595

Richter JD (2007) CPEB: a life in translation. Trends Biochem Sci 32:279–285

Robinson BG, Frim DM, Schwartz WJ, Majzoub JA (1988) Vasopressin mRNA in the suprachiasmatic nuclei: daily regulation of polyadenylate tail length. Science 241:342–344

Rudic RD, McNamara P, Curtis AM, Boston RC, Panda S, Hogenesch JB, Fitzgerald GA (2004) BMAL1 and CLOCK, two essential components of the circadian clock, are involved in glucose homeostasis. PLoS Biol 2:e377

Shimba S, Ishii N, Ohta Y, Ohno T, Watabe Y, Hayashi M, Wada T, Aoyagi T, Tezuka M (2005) Brain and muscle Arnt-like protein-1 (BMAL1), a component of the molecular clock, regulates adipogenesis. Proc Natl Acad Sci USA 102:12071–12076

Srinivasan V, Pandi-Perumal SR, Trakht I, Spence DW, Hardeland R, Poeggeler B, Cardinali DP (2009) Pathophysiology of depression: role of sleep and the melatonergic system. Psychiatr Res 165:201–214

Storch K, Lipan O, Leykin I, Viswanathan N, Davis F, Wong W, Weitz C (2002) Extensive and divergent circadian gene expression in liver and heart. Nature 417:78–83

Stubblefield JJ, Terrien J, Green CB (2012) Nocturnin: at the crossroads of clocks and metabolism. Trends Endocrinol Metab 23:326–333

Takahashi JS, Hong HK, Ko CH, McDearmon EL (2008) The genetics of mammalian circadian order and disorder: implications for physiology and disease. Nat Rev Genet 9:764–775

Toh KL, Jones CR, He Y, Eide EJ, Hinz WA, Virshup DM, Ptacek LJ, Fu YH (2001) An hPer2 phosphorylation site mutation in familial advanced sleep phase syndrome. Science 291:1040–1043

Touma C, Fenzl T, Ruschel J, Palme R, Holsboer F, Kimura M, Landgraf R (2009) Rhythmicity in mice selected for extremes in stress reactivity: behavioural, endocrine and sleep changes resembling endophenotypes of major depression. PLoS One 4:e4325

Turek FW, Joshu C, Kohsaka A, Lin E, Ivanova G, McDearmon E, Laposky A, Losee-Olson S, Easton A, Jensen DR, Eckel RH, Takahashi JS, Bass J (2005) Obesity and metabolic syndrome in circadian clock mutant mice. Science 308:1043–1045

Ueda HR, Chen W, Adachi A, Wakamatsu H, Hayashi S, Takasugi T, Nagano M, Nakahama K, Suzuki Y, Sugano S, Iino M, Shigeyoshi Y, Hashimoto S (2002) A transcription factor response element for gene expression during circadian night. Nature 418:534–539

Wagner E, Clement SL, Lykke-Andersen J (2007) An unconventional human Ccr4-Caf1 deadenylase complex in nuclear cajal bodies. Mol Cell Biol 27:1686–1695

Wang Y, Osterbur DL, Megaw PL, Tosini G, Fukuhara C, Green CB, Besharse JC (2001) Rhythmic expression of nocturnin mRNA in multiple tissues of the mouse. BMC Dev Biol 1:9

Washio-Oikawa K, Nakamura T, Usui M, Yoneda M, Ezura Y, Ishikawa I, Nakashima K, Noda T, Yamamoto T, Noda M (2007) Cnot7-null mice exhibit high bone mass phenotype and modulation of BMP actions. J Bone Mine Res 22:1217–1223

Welsh DK, Yoo SH, Liu AC, Takahashi JS, Kay SA (2004) Bioluminescence imaging of individual fibroblasts reveals persistent, independently phased circadian rhythms of clock gene expression. Curr Biol 14:2289–2295

Yamashita A, Chang TC, Yamashita Y, Zhu W, Zhong Z, Chen CY, Shyu AB (2005) Concerted action of poly(A) nucleases and decapping enzyme in mammalian mRNA turnover. Nat Struct Mol Biol 12:1054–1063

Redox and Metabolic Oscillations in the Clockwork

Akhilesh B. Reddy

Abstract Daily (circadian) clocks have evolved to coordinate behaviour and physiology around the 24-h day. Most models of the eukaryotic circadian oscillator have focused principally on transcription/translation feedback loop (TTFL) mechanisms, with accessory cytosolic loops that connect them to cellular physiology. Recent work, however, questions the absolute necessity of transcription-based oscillators for circadian rhythmicity. The recent discovery of reduction-oxidation cycles of peroxiredoxin proteins, which persist even in the absence of transcription, have prompted a reappraisal of current clock models in disparate organisms. A novel mechanism based on metabolic cycles may underlie circadian transcriptional and cytosolic rhythms, making it difficult to know where one oscillation ends and the other begins.

Introduction

Daily biological clocks provide living organisms with temporal organisation over a 24-h timescale. Organisms from bacteria to humans have evolved these rhythms to adapt their physiology to the solar cycle and anticipate the availability of resources (e.g., food and light; Bass 2012). Despite their presence in evolutionarily disparate organisms, the molecules underlying the clockwork seemed to be different in these organisms. This finding has given rise to the identification of "clock genes" that oscillate with 24-h periods but that are not related in their DNA or protein sequences, except in some cases when comparing fruit flies and mammals. This has led to the notion of the divergent evolution of different clock circuits in various model organisms, the only link being the way in which the components are joined together in a negative feedback loop topology (Rosbash 2009; Fig. 1).

A.B. Reddy (✉)
Department of Clinical Neurosciences, University of Cambridge Metabolic Research Laboratories, Cambridge, UK

NIHR Biomedical Research Centre, Wellcome Trust-MRC Institute of Metabolic Science, University of Cambridge, Addenbrooke's Hospital, Cambridge CB2 0QQ, UK
e-mail: areddy@cantab.net

P. Sassone-Corsi, Y. Christen (eds.), *A Time for Metabolism and Hormones*,
Research and Perspectives in Endocrine Interactions,
DOI 10.1007/978-3-319-27069-2_6

51

Model Organism

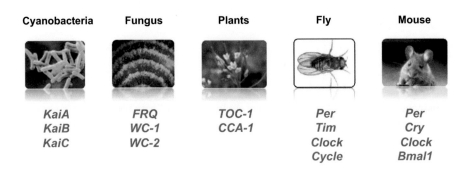

Cyanobacteria	Fungus	Plants	Fly	Mouse
KaiA	FRQ	TOC-1	Per	Per
KaiB	WC-1	CCA-1	Tim	Cry
KaiC	WC-2		Clock	Clock
			Cycle	Bmal1

" Clock Genes"

Fig. 1 The dominant model organisms used for the study of circadian rhythms are shown. Below are lists of the more important clock genes that are/were thought to mediate negative feedback loops in these organisms. At the DNA or protein sequence levels, there is no evolutionary conversation beyond some in fly and mammals

Challenging Transcriptional Models of the Clockwork

An increasing number of studies, both in higher and lower organisms, have questioned the necessity of a functional transcriptional oscillator for cellular rhythmicity. Circadian transcription is stochastic in mammalian cells (Suter et al. 2011); therefore, one would expect that perturbation of transcription during cell division would result in phase variability, which is not seen. In line with this finding, global inhibition of transcription with actinomycin D and α-amanitin has revealed the robustness of circadian oscillators to such severe perturbations, and single cells exhibit bioluminescence rhythms even when the transcription rate is reduced by ~70 % (Dibner et al. 2009).

Perhaps more importantly, studies showing that constitutive expression, or deletion, of "clock genes" does not abolish circadian rhythms call into question the importance of transcription in current clock models. In flies, expression of both *per* and *tim* under the control of a constitutive promoter can affect circadian rhythms. However, ~50 % of the flies still exhibited robust behavioral rhythms (Yang and Sehgal 2001). Similarly, fungi can exhibit conidiation (spore formation) rhythms in the absence of central components of their feedback loop [involving the *frequency* (*frq*) gene] (Lakin-Thomas 2006). Moreover, in some organisms, the dominant mechanism regulating circadian rhythms seems to be post-transcriptional, as exemplified by the circadian control of translation of luciferin binding protein (LBP) in the unicellular alga *Gonyaulax polyedra* (Morse et al. 1989; Mittag et al. 1994).

In mammals, the situation is difficult to dissect since circadian genes often have multiple homologues; therefore, double-mutant animals are generally needed to observe a behavioral phenotype. *Bmal1* was thought to be the only exception to this, with its suppression leading to clear behavioural arrhythmicity (Ko et al. 2006). Constitutive brain-specific expression of *Bmal1* in knock-out animals is, however, able to restore behavioral rhythmicity, questioning the necessity of rhythmic *Bmal1* transcription (McDearmon et al. 2006). In addition, brain-specific knockout of *Bmal1* expression produces gross pathology, with a striking abundance of activated microglia in the brains of mice, which gets progressively worse over the first 6 months of life. This finding makes it extremely difficult to dissociate the effects if BMAL1 as a generically important transcription factor from those specifically related to the malfunctioning of a biological clock (Musiek et al. 2013).

Even more importantly, imaging of suprachiasmatic nucleus (SCN) slices from arrhythmic Bmal1$-/-$ and Cry1$-/-$Cry2$-/-$ animals with bioluminescence reporters revealed the persistence of low amplitude rhythms in individual neurons (Ko et al. 2010; Maywood et al. 2011). As recently shown, it is likely that developmental effects underlie the apparent arrhythmicity that is observed when adult animals are assayed, as is the case in most experimental paradigms (Ono et al. 2013).

There is thus considerable evidence that current transcription-translation feedback loops cannot account for the multiple lines of experimental evidence that have revealed circadian oscillations in the presence of inactivated feedback loops or indeed in their absence.

Non-transcriptional Clock Mechanisms

The experimental anomalies highlighted above suggest that other mechanisms are required to fully explain the molecular basis of circadian timekeeping. It is indeed worth underscoring that transcriptional mechanisms were regarded as only one of the several possibilities that were investigated before the discovery of "clock genes" (Edmunds 1988).

An instructive outlier in clock research is the macroscopic unicellular alga *Acetabularia*, which can maintain self-sustained circadian rhythms in photosynthetic activity when its nucleus is removed by cutting off its nucleus-containing rhizoid process (Sweeney and Haxo 1961). Intriguingly, its nucleus is able to dictate the phase of oscillation but is dispensable for entrainment and phase shifting (Schweiger et al. 1964). Moreover, inhibition of transcription with Actinomycin D did not suppress rhythms in either nucleated or enucleated *Acetabularia* cells, although the former surprisingly lost rhythmicity after 2 weeks under these conditions (Mergenhagen and Schweiger 1975). Similarly, platelets were used to show that glutathione exhibited circadian oscillations relying on de novo synthesis of this important cellular reductant (Radha et al. 1985), again in the absence of a nucleus. These examples point to the fact that current circadian models cannot explain issues raised almost 40 years ago, in some cases, suggesting the existence of non-transcriptional rhythms.

How can we reconcile these seemingly opposite views? One way is to view transcription and translation in the current models as having limited roles in setting the pace of the oscillator and to note that they are needed to maintain the levels of clock proteins and to control circadian output functions. Accordingly, post-translational modifications of known clock proteins could be the fundamental oscillator, but the transcriptional oscillator would be important for robustness and could amplify post-translational oscillations. In fact, such a model exists in cyanobacteria, in which the master transcriptional regulator KaiC is part of its post-translational oscillator.

An alternative point of view is that circadian timekeeping might have evolved more than one clock in the cell to meet the requirements of precision, robustness and stability. In this case, the known transcriptional oscillator would be coupled to a post-translational oscillator. Post-translational modifications are an integral feature of the current transcription-translation feedback models, but a definitive post-translational oscillator has not yet been identified in eukaryotic species. The recent discovery of oxidation cycles in peroxiredoxin proteins (PRDXs) offers a new window on non-transcriptional rhythms in higher organisms (O'Neill and Reddy 2011; O'Neill et al. 2011; Edgar et al. 2012; Olmedo et al. 2012). More importantly, this finding immediately suggests a common phylogenetic origin for circadian timekeeping mechanisms in virtually all species relying on oxygen for energy metabolism (Edgar et al. 2012).

PRDX Rhythms

PRDXs are an antioxidant protein family involved in hydrogen peroxide metabolism and signalling (Hall et al. 2009). Their catalytic mechanism involves the oxidation of a catalytic cysteine residue in the enzymes' active site to sulfenic acid (Cys-SOH), which then forms a disulfide bond with another non-catalytic (and so-called 'resolving') cysteine residue. The thioredoxin system usually completes the cycle by reducing this disulfide bond while oxidising a molecule of NADPH. This catalytic loop has rapid turnover and allows the maintenance of low levels of intracellular hydrogen peroxide.

So-called 'typical 2-Cys PRDXs,' a subclass of PRDXs whose basic functional unit is a homodimer in which catalytic and resolving cysteine residues belong to different molecules of PRDX, are the main players implicated in circadian cycles. These can undergo further oxidation of their catalytic cysteine to sulfinic and sulfonic acid forms (Cys-SO$_{2/3}$H). The 'over-oxidised' Cys-SO$_2$H residues can be slowly recycled through adenosine triphosphate (ATP)-dependent reduction by sulfiredoxin (Rhee et al. 2007), whereas further oxidation to Cys-SO$_3$H (termed 'hyper-oxidation') is thought to be irreversible.

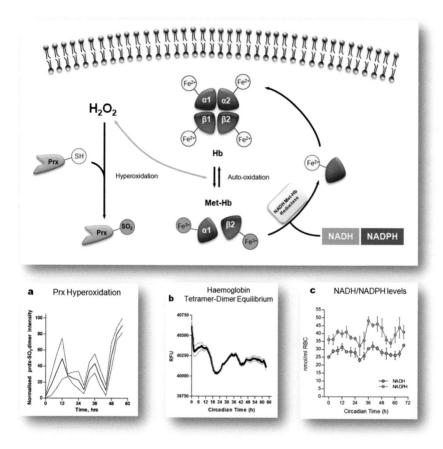

Fig. 2 A range of circadian oscillations in human red blood cells. (**a**) Oxidation of PRDXs occurs on a circadian basis in cells maintained in constant conditions (in the absence of external temporal cues) for at least 3 consecutive days. (**b**) Dynamic changes in the equilibrium of haemoglobin forms (tetramer vs. dimer states) oscillate according to a 24-h rhythm in vitro. (**c**) Oscillation of the key cellular reductants, NADH and NADPH, in red blood cells

Human red blood cells do not undergo transcription since they lack nuclei in their mature form. In these unique cells, PRDXs exhibit circadian accumulation of their dimeric over-oxidised form (PRDX-SO$_2$H) over several days (O'Neill and Reddy 2011). Such rhythms fulfil all criteria for circadian rhythms: (1) persistence in constant conditions; (2) the ability to be entrained (via temperature cycles in this case); and (3) temperature compensation (the clock does not run faster in higher temperatures). In addition, these redox rhythms are accompanied by oscillations in haemoglobin oxidation and metabolic variables, including NADH and NADPH (Fig. 2). Similar results have also recently been found in mouse red blood cells (Cho et al. 2014).

Rhythms similar to these are also present in the unicellular alga *Ostreococcus tauri*, even when transcription is inhibited by prolonged darkness (O'Neill et al. 2011); they are autophototrophic, requiring light for synthesis of most cellular substrates including RNA. Moreover, the deep phylogenetic conservation of PRDX redox rhythms extends to include fungal, plant, bacterial and even archaeal species. Critically, such rhythms are not dependent on previously identified clock genes, since mutants lacking circadian components maintain redox oscillations, albeit slightly phase-shifted (Edgar et al. 2012).

The phylogenetic conservation of PRDX rhythms suggests that primordial redox oscillators probably evolved following the Great Oxidation Event 2.5 billion years ago. At this time, photosynthetic bacteria are thought to have acquired the ability to produce oxygen from water, which caused a dramatic rise in Earth's atmospheric oxygen. Rhythmic production of oxygen and reactive oxygen species (ROS) by sunlight may therefore have been a critical driving force in the co-evolution of clock mechanisms and ROS removal systems that could anticipate, and thus resonate with, externally driven redox cycles (Bass 2012; Edgar et al. 2012; Fig. 3).

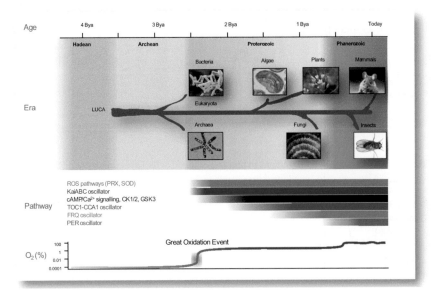

Fig. 3 Phylogenetic origins of circadian oscillatory systems. A timeline is shown at the *top* of the schematic, with the geological era illustrated. A schematic phylogenetic tree shows the origins of each organism studied, stemming from the last universal common ancestor (LUCA). The putative epoch over which each oscillator system has existed is illustrated by the labeled bars. CK1/2, casein kinase 1 or 2; GSK3, glycogen synthase kinase 3; SOD, superoxide dismutase (Adapted from Edgar et al. 2012)

Redox and Metabolic Clocks in Eukaryotes

There is clearly an interplay between circadian and metabolic cycles, and there is good evidence of reciprocal effects that disruption of one cycle has on the other at physiological and molecular levels (Bass 2012; Rey and Reddy 2013). High-fat diet, for example, lengthens the behavioral period of rhythms in mice and changes the expression pattern of clock genes (Kohsaka et al. 2007). Conversely, healthy patients subjected to 3 weeks of circadian disruption exhibit pre-diabetic symptoms (Buxton et al. 2012). The growing evidence suggesting that circadian rhythms are fundamentally metabolic requires that currently understood transcriptional oscillations are tightly coupled to metabolic cycles. This hypothesis is strongly supported by the numerous examples of accessory loops embedding the circadian transcriptional clock within cellular metabolism (Fig. 4).

An accessory loop involving $NAD^+/NADH$ is likely to play an important role in connecting cytosolic and compartment-specific redox states to transcriptional clock components such as PER2 (Asher et al. 2008) and CLOCK/BMAL1 (Rutter et al. 2001; Nakahata et al. 2008; Asher et al. 2010; Yoshii et al. 2013). In addition, other redox-sensitive mechanisms have been identified in the clockwork and, in particular, the heme-sensing transcriptional regulators (Dioum et al. 2002; Yin et al. 2007; Gupta et al. 2011).

Even in early molecular studies of the circadian clock, before "clock genes" had been discovered in any model organism, rhythms in redox had been reported. For example, in plants, $NADP^+$:NADPH ratio exhibited circadian cycles in seedlings kept in constant darkness (Wagner and Frosch 1974). Several studies in rodents

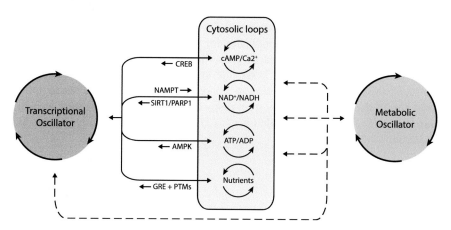

Fig. 4 Links between transcriptional, cytosolic, and metabolic cycles. Cytosolic processes are thought to be part of the transcription/translation feedback loop (TTFL). The latter are involved mainly in redox and energy metabolism and form accessory loops that are controlled by the TTFL oscillator and, in turn, feed back to it. Abbreviations: *CREB* cAMP response element-binding protein, *GRE* glucose response element, *PARP* poly(ADP-ribose) polymerase, *PTMs* posttranslational modifications (Adapted from Reddy and Rey 2014)

showed that redox parameters, including the glutathione redox ratio, were diurnally regulated in the liver, although it is possible that these oscillations might have been partially driven by food intake (Isaacs and Binkley 1977a, b; Robinson et al. 1981; Kaminsky et al. 1984; Belanger et al. 1991). Nevertheless, human platelets kept in vitro showed circadian rhythms in glutathione content (Radha et al. 1985), suggesting that feeding cycles might internally resonate with these cell-autonomous biochemical rhythms.

The hypothesis that metabolic cycles might be a fundamental mechanism underlying biological clocks has been proposed based on both theoretical and experimental observations (Roenneberg and Merrow 1999). Potential evidence for this hypothesis in mammals has come from the McKnight and Sagami groups, who have shown that BMAL1/CLOCK DNA-binding activity can be modulated in vitro by the redox poise of $NAD(P)^+/NAD(P)H$ coenzymes (Rutter et al. 2001; Yoshii et al. 2013). In addition, the action of BMAL1/CLOCK on the NAD^+-producing enzymes lactate dehydrogenase (*Ldh*) and *Nampt* could potentially feed back onto intracellular redox balance (Rutter et al. 2001; Nakahata et al. 2009; Ramsey et al. 2009). These results still require in vivo confirmation, given the relatively high concentration (millimolar range) of the coenzymes used in in vitro assays previously (Rutter et al. 2001; Yoshii et al. 2013). The recent discovery of PRDX oscillations in non-transcriptional systems, however, offers supportive evidence that redox cycles can function as circadian oscillators in their own right.

It is evident that, in organisms in which metabolic oscillations have been found but transcription-translation feedback loops have not, as in the worm *Caenorhabditis elegans*, insights into metabolic oscillatory mechanisms may be easier to come by. It is thus possible that metabolic oscillations could drive PRDX oscillations in the absence of known transcriptional feedback oscillators (Olmedo et al. 2012). So-called accessory loops, including $NAD^+/NADH$ and $NADP^+/NADPH$ cycles, are potential candidates for self-sustained metabolic oscillators, but further studies of their oscillatory properties in clock mutant backgrounds will be of great interest to identify bona fide components of metabolic oscillators. However, this assumes that deletion of important circadian-relevant transcription factors itself does not lead to abhorrent redox changes in cells and tissues, as is the case in *Bmal1*$-/-$ animals (Kondratov et al. 2006), which could compromise redox oscillations indirectly.

Conclusion

Metabolic non-transcriptional cycles clearly interlock with transcriptional processes in the circadian system. The peroxiredoxin system could be part of an uncharacterised metabolic oscillator, given its broad phylogenetic conservation and its slow kinetics, which is compatible with 24-h rhythmicity. Establishing the molecular links between fundamental cellular redox metabolism and transcriptional components of the clockwork remains an exciting challenge in the field.

References

Asher G, Gatfield D, Stratmann M, Reinke H, Dibner C, Kreppel F, Mostoslavsky R, Alt FW, Schibler U (2008) SIRT1 regulates circadian clock gene expression through PER2 deacetylation. Cell 134:317–328

Asher G, Reinke H, Altmeyer M, Gutierrez-Arcelus M, Hottiger MO, Schibler U (2010) Poly(ADP-ribose) polymerase 1 participates in the phase entrainment of circadian clocks to feeding. Cell 142:943–953

Bass J (2012) Circadian topology of metabolism. Nature 491:348–356

Belanger PM, Desgagné M, Bruguerolle B (1991) Temporal variations in microsomal lipid peroxidation and in glutathione concentration of rat liver. Drug Metab Dispos 19:241–244, Available at: http://eutils.ncbi.nlm.nih.gov/entrez/eutils/elink.fcgi?dbfrom=pubmed&id=1673408&retmode=ref&cmd=prlinks

Buxton OM, Cain SW, O'Connor SP, Porter JH, Duffy JF, Wang W, Czeisler CA, Shea SA (2012) Adverse metabolic consequences in humans of prolonged sleep restriction combined with circadian disruption. Sci Translat Med 4:129ra43

Cho CS, Yoon HJ, Kim JY, Woo HA, Rhee SG (2014) Circadian rhythm of hyperoxidized peroxiredoxin II is determined by hemoglobin autoxidation and the 20S proteasome in red blood cells. Proc Natl Acad Sci USA 111:12043–12048

Dibner C, Sage D, Unser M, Bauer C, d'Eysmond T, Naef F, Schibler U (2009) Circadian gene expression is resilient to large fluctuations in overall transcription rates. EMBO J 28:123–134

Dioum EM, Dioum EM, Rutter J, Tuckerman JR, Gonzalez G, Gilles-Gonzalez M-A, McKnight SL (2002) NPAS2: a gas-responsive transcription factor. Science 298:2385–2387

Edgar RS, Green EW, Zhao Y, van Ooijen G, Olmedo M, Qin X, Xu Y, Pan M, Valekunja UK, Feeney KA, Maywood ES, Hastings MH, Baliga NS, Merrow M, Millar AJ, Johnson CH, Kyriacou CP, O'Neill JS, Reddy AB (2012) Peroxiredoxins are conserved markers of circadian rhythms. Nature 485:459–464

Edmunds LNJ (1988) Cellular and molecular bases of biological clocks. Springer, New York, p 514

Gupta N, Gupta N, Ragsdale SW, Ragsdale SW (2011) Thiol-disulfide redox dependence of heme binding and heme ligand switching in nuclear hormone receptor rev-erb. J Biol Chem 286:4392–4403

Hall A, Karplus PA, Poole LB (2009) Typical 2-Cys peroxiredoxins—structures, mechanisms and functions. FEBS J 276:2469–2477

Isaacs J, Binkley F (1977a) Glutathione dependent control of protein disulfide-sulfhydryl content by subcellular fractions of hepatic tissue. Biochim Biophys Acta 497:192–204, Available at: http://eutils.ncbi.nlm.nih.gov/entrez/eutils/elink.fcgi?dbfrom=pubmed&id=557349&retmode=ref&cmd=prlinks

Isaacs JT, Binkley F (1977b) Cyclic AMP-dependent control of the rat hepatic glutathione disulfide-sulfhydryl ratio. Biochim Biophys Acta 498:29–38, Available at: http://eutils.ncbi.nlm.nih.gov/entrez/eutils/elink.fcgi?dbfrom=pubmed&id=18207&retmode=ref&cmd=prlinks

Kaminsky YG, Kosenko EA, Kondrashova MN (1984) Analysis of the circadian rhythm in energy metabolism of rat liver. Int J Biochem 16:629–639, Available at: http://eutils.ncbi.nlm.nih.gov/entrez/eutils/elink.fcgi?dbfrom=pubmed&id=6468728&retmode=ref&cmd=prlinks

Ko CH, Ko CH, Takahashi JS (2006) Molecular components of the mammalian circadian clock. Hum Mol Genet 15:R271–R277

Ko CH, Yamada YR, Welsh DK, Buhr ED, Liu AC, Zhang EE, Ralph MR, Kay SA, Forger DB, Takahashi JS (2010) Emergence of noise-induced oscillations in the central circadian pacemaker. PLoS Biol 8:e1000513

Kohsaka A, Laposky AD, Ramsey KM, Estrada C, Joshu C, Kobayashi Y, Turek FW, Bass J (2007) High-fat diet disrupts behavioral and molecular circadian rhythms in mice. Cell Metab 6:414–421

Kondratov RV, Kondratova AA, Gorbacheva VY, Vykhovanets OV, Antoch MP (2006) Early aging and age-related pathologies in mice deficient in BMAL1, the core component of the circadian clock. Genes Dev 20:1868–1873

Lakin-Thomas PL (2006) Transcriptional feedback oscillators: maybe, maybe not. J Biol Rhythms 21:83–92

Maywood ES, Chesham JE, O'Brien JA, Hastings MH (2011) A diversity of paracrine signals sustains molecular circadian cycling in suprachiasmatic nucleus circuits. Proc Natl Acad Sci USA 108:14306–14311

McDearmon EL, Patel KN, Ko CH, Walisser JA, Schook AC, Chong JL, Wilsbacher LD, Song EJ, Hong H-K, Bradfield CA, Takahashi JS (2006) Dissecting the functions of the mammalian clock protein BMAL1 by tissue-specific rescue in mice. Science 314:1304–1308

Mergenhagen D, Schweiger HG (1975) The effect of different inhibitors of transcription and translation on the expression and control of circadian rhythm in individual cells of Acetabularia. Exp Cell Res 94:321–326, Available at: http://eutils.ncbi.nlm.nih.gov/entrez/eutils/elink.fcgi?dbfrom=pubmed&id=1193133&retmode=ref&cmd=prlinks

Mittag M, Lee DH, Hastings JW (1994) Circadian expression of the luciferin-binding protein correlates with the binding of a protein to the 3′ untranslated region of its mRNA. Proc Natl Acad Sci USA 91:5257–5261

Morse D, Milos PM, Roux E, Hastings JW (1989) Circadian regulation of bioluminescence in Gonyaulax involves translational control. Proc Natl Acad Sci USA 86:172–176

Musiek ES, Lim MM, Yang G, Bauer AQ, Qi L, Lee Y, Roh JH, Ortiz-Gonzalez X, Dearborn JT, Culver JP, Herzog ED, Hogenesch JB, Wozniak DF, Dikranian K, Giasson BI, Weaver DR, Holtzman DM, Fitzgerald GA (2013) Circadian clock proteins regulate neuronal redox homeostasis and neurodegeneration. J Clin Invest 123:5389–5400

Nakahata Y, Kaluzova M, Grimaldi B, Sahar S, Hirayama J, Chen D, Guarente LP, Sassone-Corsi P (2008) The NAD+-dependent deacetylase SIRT1 modulates CLOCK-mediated chromatin remodeling and circadian control. Cell 134:329–340

Nakahata Y, Sahar S, Astarita G, Kaluzova M, Sassone-Corsi P (2009) Circadian control of the NAD+ salvage pathway by CLOCK-SIRT1. Science 324:654–657

O'Neill JS, Reddy AB (2011) Circadian clocks in human red blood cells. Nature 469:498–503

O'Neill JS, van Ooijen G, Dixon LE, Troein C, Corellou F, Bouget F-Y, Reddy AB, Millar AJ (2011) Circadian rhythms persist without transcription in a eukaryote. Nature 469:554–558

Olmedo M, O'Neill JS, Edgar RS, Valekunja UK, Reddy AB, Merrow M (2012) Circadian regulation of olfaction and an evolutionarily conserved, nontranscriptional marker in Caenorhabditis elegans. Proc Natl Acad Sci USA 109:20479–20484

Ono D, Honma S, Honma K-I (2013) Cryptochromes are critical for the development of coherent circadian rhythms in the mouse suprachiasmatic nucleus. Nat Commun 4:1666

Radha E, Hill TD, Rao GH, White JG (1985) Glutathione levels in human platelets display a circadian rhythm in vitro. Thromb Res 40:823–831, Available at: http://eutils.ncbi.nlm.nih.gov/entrez/eutils/elink.fcgi?dbfrom=pubmed&id=4089839&retmode=ref&cmd=prlinks

Ramsey KM, Yoshino J, Brace CS, Abrassart D, Kobayashi Y, Marcheva B, Hong HK, Chong JL, Buhr ED, Lee C, Takahashi JS, Imai SI, Bass J (2009) Circadian clock feedback cycle through NAMPT-mediated NAD+ biosynthesis. Science 324:651–654

Reddy AB, Rey G (2014) Metabolic and nontranscriptional circadian clocks: eukaryotes. Annu Rev Biochem 83:165–189

Rey G, Reddy AB (2013) Connecting cellular metabolism to circadian clocks. Trends Cell Biol 23: 234–241

Rhee SG, Jeong W, Chang TS, Woo HA (2007) Sulfiredoxin, the cysteine sulfinic acid reductase specific to 2-Cys peroxiredoxin: its discovery, mechanism of action, and biological significance. Kidney Int Suppl 106:S3–S8

Robinson JL, Foustock S, Chanez M, Bois-Joyeux B, Peret J (1981) Circadian variation of liver metabolites and amino acids in rats adapted to a high protein, carbohydrate-free diet. J Nutr 111:1711–1720, Available at: http://eutils.ncbi.nlm.nih.gov/entrez/eutils/elink.fcgi?dbfrom=pubmed&id=7288496&retmode=ref&cmd=prlinks

Roenneberg T, Merrow M (1999) Circadian systems and metabolism. J Biol Rhythms 14:449–459

Rosbash M (2009) The implications of multiple circadian clock origins. PLoS Biol 7:e62

Rutter J, Reick M, Wu LC, McKnight SL (2001) Regulation of clock and NPAS2 DNA binding by the redox state of NAD cofactors. Science 293:510–514. doi:10.1126/science.1060698

Schweiger E, Wallraff HG, Schweiger HG (1964) Endogenous circadian rhythm in cytoplasm of Acetabularia: influence of the nucleus. Science 146:658–659

Suter DM, Suter DM, Molina N, Molina N, Gatfield D, Gatfield D, Schneider K, Schneider K, Schibler U, Schibler U, Naef F, Naef F (2011) Mammalian genes are transcribed with widely different bursting kinetics. Science 332:472–474

Sweeney BM, Haxo FT (1961) Persistence of a photosynthetic rhythm in enucleated Acetabularia. Science 134:1361–1363

Wagner E, Frosch S (1974) Cycles in plants. J Interdis Cycle Res 5:231–239

Yang Z, Sehgal A (2001) Role of molecular oscillations in generating behavioral rhythms in Drosophila. Neuron 29:453–467

Yin L, Wu N, Curtin JC, Qatanani M, Szwergold NR, Reid RA, Waitt GM, Parks DJ, Pearce KH, Wisely GB, Lazar MA (2007) Rev-erbalpha, a heme sensor that coordinates metabolic and circadian pathways. Science 318:1786–1789

Yoshii K, Ishijima S, Sagami I (2013) Effects of NAD(P)H and its derivatives on the DNA-binding activity of NPAS2, a mammalian circadian transcription factor. Biochem Biophys Res Commun 437:386–391

Rev-erbs: Integrating Metabolism Around the Clock

Mitchell A. Lazar

Abstract Mammalian circadian and metabolic physiologies are intertwined, and the nuclear Rev-erbα is a key transcriptional link between them. Rev-erbα, and the highly related Rev-erbβ, are potent transcriptional repressors that are required for the function of the core mammalian molecular clock. The Rev-erbs are also critical regulators of clock output in metabolic cells and tissues. This chapter focuses on the physiological functions of Rev-erbα and β in regulating circadian rhythms and metabolism in mammalian tissues.

Introduction

Much of biology is conducted with rhythms that have a phase of approximately 24 h, matching the duration of a day on planet Earth (Huang et al. 2011). The genetic basis of these circadian rhythms was unveiled in the fruit fly, *Drosophila melanogaster*, where the clock mechanism involves feedback regulation by factors whose own expression exhibit circadian rhythmicity (Rosbash et al. 1996). These factors function as transcriptional regulators, and it is now recognized that most genomes, including those of all mammals that have been evaluated, are transcribed in a rhythmic manner (Schibler 2006).

The mammalian clock mechanism involves interconnected transcriptional and translational feedback loops, where the most well-understood positive regulator is a heterodimer of the basic helix-loop-helix (HLH) transcription factors BMAL1 and CLOCK (King and Takahashi 2000). In addition to positively regulating clock output genes, the BMAL1/CLOCK heterodimer activates the expression of two negative regulators. One is another bHLH heterodimer, comprised of the proteins PERIOD (PER) and CRYPTOCHROME (CRY), which interact with BMAL1/CLOCK to

M.A. Lazar, M.D., Ph.D. (✉)
Division of Endocrinology, Diabetes, and Metabolism, Department of Medicine, Department of Genetics, and The Institute for Diabetes, Obesity, and Metabolism, Perelman School of Medicine at the University of Pennsylvania, Philadelphia, PA 19104, USA
e-mail: lazar@mail.med.upenn.edu

© The Author(s) 2016
P. Sassone-Corsi, Y. Christen (eds.), *A Time for Metabolism and Hormones*,
Research and Perspectives in Endocrine Interactions,
DOI 10.1007/978-3-319-27069-2_7

interfere with its activity (King and Takahashi 2000). The second repressive loop is mediated by the Rev-erb nuclear receptors (NRs) α and β, of which Rev-erbα is the more highly functional (Everett and Lazar 2014). This chapter will focus on the Rev-erbs, particularly on the more well-studied Rev-erbα.

Repression of Transcription by Rev-erbs

Rev-erbs belong to a large NR superfamily of ligand-regulated transcription factors (Evans 2013). Discovered in 1989 (Lazar et al. 1989; Miyajima et al. 1989), Rev-erbα was one of the first identified orphan NRs, i.e., a member of the family whose ligand was not predicted from earlier physiology and biochemical studies (Mullican et al. 2013). The highly related Rev-erbβ was identified in 1994 (Bonnelye et al. 1994; Dumas et al. 1994; Forman et al. 1994; Retnakaran et al. 1994). Molecular heme has been identified as the endogenous ligand for Rev-erbα and Rev-erbβ (Raghuram et al. 2007; Yin et al. 2007). Although the physiological function of this regulation is not well understood, the ability to sense heme levels may position Rev-erb as a mediator of metabolic effects on metabolism.

Rev-erbs bind sequence-specifically to DNA, with the preferred binding site consisting of the classical NR half-site AGGTCA flanked by an A/T-rich 5′ sequence (Harding and Lazar 1993). This binding site is referred to as the RevRE or as the RORE, as it is also bound by the Retinoic Acid Receptor-related Orphan Receptor (ROR; Giguere et al. 1994). The DNA-binding domain (DBD) of Rev-erbα binds in the major groove of the AGGTCA half-site, whereas a C-terminal extension makes minor groove contacts with the A/T-rich 5′ sequence (Zhao et al. 1998). Rev-erbs bind as a monomer to this site but bind even more tightly as a dimer to a direct repeat with a 2 base pair spacer, referred to as the RevDR2 (Harding and Lazar 1995).

Rev-erbs lack the C-terminal region that is required for ligand-dependent transcriptional activation by other NRs (Glass and Rosenfeld 2000). Thus, they function primarily as potent repressors of transcription when bound to DNA (Zamir et al. 1997), interacting constitutively with the Nuclear Corepressor 1 (NCoR; Horlein et al. 1995; Zamir et al. 1996). NCoR is a large protein (~270 kDa) with inherent repressive function as well as several short helical domains that specifically interact with NRs, called the corepressor-NR (CoRNR) boxes (Hu and Lazar 1999). Heme further stabilizes its interaction with full-length, endogenous NCoR (Raghuram et al. 2007; Yin et al. 2007). In addition to serving as a heme sensor, the Rev-erb activity may also be sensitive to the oxidation state of the heme iron (Marvin et al. 2009). To bind NCoR stably enough to actively repress transcription, two Rev-erbα molecules must interact with CoRNR peptides from NCoR; this interaction can occur at the RevDR2 site, which the Rev-erbs bind cooperatively as a dimer, or at two Revre/RORE sites bound independently by Rev-erb monomers (Zamir et al. 1997).

NCoR represses transcription by nucleating a large multiprotein repressor complex, which impacts the epigenome and the function of core transcriptional factors and RNA polymerase II (Guenther et al. 2000; Yoon et al. 2003). Stoichiometric components of the NCoR complex include Transducin Beta-Like 1 (TBL1), G-protein Pathway Suppressor 2 (GPS2), and Histone Deacetylase 3 (HDAC3; Guenther et al. 2000; Zhang et al. 2002; Yoon et al. 2003). HDAC3 is of particular interest, because it is an epigenomic modulator that deacetylates lysine residues in the tails of nucleosomal histone proteins to create a repressive chromatin environment (Haberland et al. 2009). NCoR and HDAC3 are both required for Rev-erbα to repress *Bmal1* gene transcription (Yin and Lazar 2005), and both NCoR and HDAC3 are associated with Rev-erbα at thousands of DNA binding sites genome wide in the mouse liver (Feng et al. 2011).

Circadian Biology of Rev-erbs

In 1998, Rev-erbα was noted to one of the genes that oscillates within the circadian transcriptome of mammalian cells cycling in tissue culture (Balsalobre et al. 1998). In mice Rev-erbα mRNA expression is robustly circadian in multiple tissues (Yang et al. 2006), and genetic deletion of Rev-erbα shortens the period of behavioral rhythms by ~30 min in the absence of daily light cues (Preitner et al. 2002). Rev-erbα modulates the rhythmicity of additional circadian regulators, including *Clock* (Crumbley and Burris 2011), *Cry1* (Ukai-Tadenuma et al. 2011), *Nfil3/ E4Bp4* (Duez et al. 2008), and *Npas2* (Crumbley et al. 2010) and thus has a major influence on the cell-autonomous molecular timing system. Indeed, constitutive expression of Rev-erbα in mouse liver represses the majority of cycling transcripts (Kornmann et al. 2006). Importantly, ablation of both Rev-erbα and β abrogates circadian gene expression in mouse embryonic fibroblasts, demonstrating a fundamental requirement for the Rev-erbs (Bugge et al. 2012). Moreover, genetic mutation of Rev-erbα and β caused arrhythmic behavior in mice (Cho et al. 2012). Therefore Rev-erbα and β are both required components of the core clock machinery. Loss of either Rev-erb alone is insufficient to abolish circadian rhythms, indicating that their clock functions are redundant, although Rev-erbα is more critical because its absence modestly disrupts normal circadian rhythms whereas the loss of Rev-erbβ does not.

Rev-erbα and Metabolism

Circadian rhythms and metabolism are highly intertwined (Eckel-Mahan and Sassone-Corsi 2013), and indeed Rev-erbα regulates metabolic function in many tissues. In the liver, Rev-erbα regulates cholesterol and bile acid metabolism (Duez et al. 2008; Le Martelot et al. 2009), and more recently has been observed to play a key role in the circadian regulation of triglyceride metabolism (Feng et al. 2011). Rev-erbα binds widely and robustly to the genome at ZT10, when its expression is maximal; however, it binds to very few sites when its expression is at a nadir, such as at ZT22. This genomic binding is enriched at genes involved in lipid metabolism and, indeed, mice lacking Rev-erbα have mild fatty liver, or hepatic steatosis (Feng et al. 2011). The oscillatory expression of Rev-erbα regulates circadian gene expression directly at target genes with strong binding motifs, whose circadian expression is antiphase to that of Rev-erbα, as well as indirectly by repression of another circadian repressor called E4BP4, whose target genes are expressed in phase with Rev-erbα (Fang et al. 2014). The liver cistrome of Rev-erbβ is quite similar, and knockdown of Rev-erbβ in livers of Rev-erbα null mice caused a more markedly fatty liver (Bugge et al. 2012). NCoR and HDAC3 bind to the genome at the vast majority of Rev-erb sites and, indeed, ablation of either NCoR or HDAC3 in mouse liver leads to marked hepatic steatosis (Knutson et al. 2008; Sun et al. 2012, 2013).

Studies of adipocyte differentiation in cultured cell lines have suggested that Rev-erbα plays an important role in adipocyte differentiation (Chawla and Lazar 1993; Fontaine et al. 2003; Wang and Lazar 2008), yet white adipose tissue (WAT) mass was not reduced in mice lacking Rev-erbα (Chomez et al. 2000; Delezie et al. 2012), indicating that Rev-erbα is not absolutely required for adipocyte formation in vivo. Rev-erbα may play a role in brown adipose tissue (BAT), which is a major site of thermogenesis (Gerhart-Hines et al. 2013). Circadian expression of Rev-erbα in BAT peaks at ZT10, which is antiphase to the circadian rhythm of body temperature. Mice lacking Rev-erbα have a higher nadir in body temperature, at least in part due to derepression of Uncoupling Protein 1 (UCP1), which is a circadian target of Rev-erbα and constitutively high in the BAT of mice genetically lacking Rev-erbα (Gerhart-Hines et al. 2013). Mice also have an increased vulnerability to cold temperature at times of day when Rev-erbα levels are high; this vulnerability is ameliorated in the absence Rev-erbα (Gerhart-Hines et al. 2013).

A role for Rev-erbα in skeletal myocytes was first identified in C2C12 cultured myoblasts, where Rev-erbα represses the expression of genes involved in muscle cell differentiation (Downes et al. 1995). Rev-erbα mRNA expression is circadian manner in mouse skeletal muscle (Yang et al. 2006), and loss of Rev-erbα function reduces mitochondrial content and function, leading to an impaired exercise capacity (Woldt et al. 2013). It should be noted that the transcriptomic changes in muscle are not observed in liver or BAT and thus reflect tissue-specific functions of Rev-erbα.

Rev-erbα is also expressed in a circadian manner in the pancreatic islets and plays a role in the function of insulin-producing β-cells and glucagon-producing α-cells (Vieira et al. 2012, 2013). Islets isolated at the peak of Rev-erbα expression have higher levels of glucose-stimulated insulin secretion (Vieira et al. 2012), and Rev-erbα also promotes glucagon secretion in α-cells of the pancreas (Vieira et al. 2013).

Inflammatory cells are increasingly linked to metabolic function (Osborn and Olefsky 2012), and Rev-erbα mediates the circadian gating of the LPS-induced endotoxic response (Gibbs et al. 2012). Genome-wide studies of Rev-erbα and Rev-erbβ cistromes and transcriptomes suggest that Rev-erbα influences macrophage gene expression at bindings sites marked by hematopoietic transcription factors, including PU.1 (Lam et al. 2013).

Conclusions

The nuclear receptor Rev-erbα acts in a tissue-specific manner to regulate circadian rhythms as well as metabolism, in some cases acting redundantly with Rev-erbβ. A critical question is whether Rev-erbα can be targeted for therapeutic purposes. Synthetic pharmacological agonists have been developed (Grant et al. 2010; Solt et al. 2012), yet the tissue-specific complexity of Rev-erb biology raises major challenges to human therapeutics. Perhaps the dramatic circadian expression of Rev-erbs can be exploited by timing drug delivery to selectively impact their specific and beneficial functions in integration of metabolism and the circadian clock.

Acknowledgments I thank members of my research group, particularly Logan Everett, for helpful discussions. Research on Rev-erb in my laboratory is supported by NIH grant R01 DK45586 (MAL).

References

Balsalobre A, Damiola F, Schibler U (1998) A serum shock induces circadian gene expression in mammalian tissue culture cells. Cell 93:929–937
Bonnelye E, Vanacker JM, Desbiens X, Begue A, Stehelin D, Laudet V (1994) Rev-erb-beta, a new member of the nuclear receptor superfamily, is expressed in the nervous-system during chicken development. Cell Growth Differ 5:1357–1365
Bugge A, Feng D, Everett LJ, Briggs ER, Mullican SE, Wang FF, Jager J, Lazar MA (2012) Rev-erb alpha and Rev-erb beta coordinately protect the circadian clock and normal metabolic function. Genes Dev 26:657–667

Chawla A, Lazar MA (1993) induction of Rev-erbA-alpha, a nuclear hormone receptor-related transcriptional activator, during adipocyte differentiation. J Biol Chem 268:16265–16269

Cho H, Zhao X, Hatori M, Yu RT, Barish GD, Lam MT, Chong LW, DiTacchio L, Atkins AR, Glass CK, Liddle C, Auwerx J, Downes M, Panda S, Evans RM (2012) Regulation of circadian behaviour and metabolism by REV-ERB-alpha and REV-ERB-beta. Nature 485:123–127

Chomez P, Neveu I, Mansen A, Kiesler E, Larsson L, Vennstrom B, Arenas E (2000) Increased cell death and delayed development in the cerebellum of mice lacking the Rev-erbA alpha orphan receptor. Development 127:1489–1498

Crumbley C, Burris TP (2011) Direct regulation of CLOCK expression by REV-ERB. PLoS One 6:e17290

Crumbley C, Wang YJ, Kojetin DJ, Burris TP (2010) Characterization of the core mammalian clock component, NPAS2, as a REV-ERB alpha/ROR alpha target gene. J Biol Chem 285:35386–35392

Delezie J, Dumont S, Dardente H, Oudart H, Grechez-Cassiau A, Klosen P, Teboul M, Delaunay F, Pevet P, Challet E (2012) The nuclear receptor REV-ERB alpha is required for the daily balance of carbohydrate and lipid metabolism. FASEB J 26:3321–3335

Downes M, Carozzi AJ, Muscat GEO (1995) Constitutive expression of the orphan receptor, Rev-erbA-alpha, inhibits muscle differentiation and abrogates the expression of the myoD gene family. Mol Endocrinol 9:1666–1678

Duez H, Van Der Veen JN, Duhem C, Pourcet B, Touvier T, Fontaine C, Derudas B, Bauge E, Havinga R, Bloks VW, Wolters H, Van Der Sluijs FH, Vennstrom B, Kuipers F, Staels B (2008) Regulation of bile acid synthesis by the nuclear receptor Rev-erb alpha. Gastroenterology 135:689–698

Dumas B, Harding HP, Choi HS, Lehmann KA, Chung M, Lazar MA, Moore DD (1994) A new orphan member of the nuclear hormone-receptor superfamily closely-related to Rev-erb. Mol Endocrinol 8:996–1005

Eckel-Mahan K, Sassone-Corsi P (2013) Metabolism and the circadian clock converge. Physiol Rev 93:107–135

Evans RM (2013) Journal of Molecular Endocrinology 25th anniversary special issue. J Mol Endocrinol 51:E1–E3

Everett LJ, Lazar MA (2014) Nuclear receptor Rev-erb alpha: up, down, and all around. Trends Endocrinol Metab 25:586–592

Fang B, Everett LJ, Jager J, Brigges E, Armour SM, Feng D, Roy A, Gerhart-Hines Z, Sun Z, Lazar MA (2014) Circadian enhancers coordinate multiple phases of rhythmic gene transcription in vivo. Cell 159:1140–1152

Feng D, Liu T, Sun Z, Bugge A, Mullican SE, Alenghat T, Liu XS, Lazar MA (2011) A circadian rhythm orchestrated by histone deacetylase 3 controls hepatic lipid metabolism. Science 331:1315–1319

Fontaine C, Dubois G, Duguay Y, Helledie T, Vu-Dac N, Gervois P, Soncin F, Mandrup S, Fruchart JC, Fruchart-Najib J, Staels B (2003) The orphan nuclear receptor Rev-Erb alpha is a peroxisome proliferator-activated receptor (PPAR) gamma target gene and promotes PPAR gamma-induced adipocyte differentiation. J Biol Chem 278:37672–37680

Forman BM, Chen J, Blumberg B, Kliewer SA, Henshaw R, Ong ES, Evans RM (1994) Cross-talk among ROR-alpha-1 and the Rev-erb family of orphan nuclear receptors. Mol Endocrinol 8:1253–1261

Gerhart-Hines Z, Feng D, Emmett MJ, Everett LJ, Loro E, Briggs ER, Bugge A, Hou C, Ferrara C, Seale P, Pryma DA, Khurana TS, Lazar MA (2013) The nuclear receptor Rev-erb alpha controls circadian thermogenic plasticity. Nature 503:410–413

Gibbs JE, Blaikley J, Beesley S, Matthews L, Simpson KD, Boyce SH, Farrow SN, Else KJ, Singh D, Ray DW, Loudon ASI (2012) The nuclear receptor REV-ERB alpha mediates circadian regulation of innate immunity through selective regulation of inflammatory cytokines. Proc Natl Acad Sci USA 109:582–587

Giguere V, Tini M, Flock G, Ong E, Evans RM, Otulakowski G (1994) Isoform-specific amino-terminal domains dictate dna-binding properties of ROR-alpha, a novel family of orphan hormone nuclear receptors. Genes Dev 8:538–553

Glass CK, Rosenfeld MG (2000) The coregulator exchange in transcriptional functions of nuclear receptors. Genes Dev 14:121–141

Grant D, Yin L, Collins JL, Parks DJ, Orband-Miller LA, Wisely GB, Joshi S, Lazar MA, Willson TM, Zuercher WJ (2010) GSK4112, a small molecule chemical probe for the cell biology of the nuclear heme receptor Rev-erb alpha. ACS Chem Biol 5:925–932

Guenther MG, Lane WS, Fischle W, Verdin E, Lazar MA, Shiekhattar R (2000) A core SMRT corepressor complex containing HDAC3 and TBL1, a WD40-repeat protein linked to deafness. Genes Dev 14:1048–1057

Haberland M, Montgomery RL, Olson EN (2009) The many roles of histone deacetylases in development and physiology: implications for disease and therapy. Nat Rev Genet 10:32–42

Harding HP, Lazar MA (1993) The orphan receptor Rev-erbA-alpha activates transcription via a novel response element. Mol Cell Biol 13:3113–3121

Harding HP, Lazar MA (1995) The monomer-binding orphan receptor Rev-erb represses transcription as a dimer on a novel direct repeat. Mol Cell Biol 15:4791–4802

Horlein AJ, Naar AM, Heinzel T, Torchia J, Gloss B, Kurokawa R, Ryan A, Kamel Y, Soderstrom M, Glass CK, Rosenfeld MG (1995) Ligand-independent repression by the thyroid-hormone receptor-mediated by a nuclear receptor co-repressor. Nature 377:397–404

Hu X, Lazar MA (1999) The CoRNR motif controls the recruitment of corepressors by nuclear hormone receptors. Nature 402:93–96

Huang WY, Ramsey KM, Marcheva B, Bass J (2011) Circadian rhythms, sleep, and metabolism. J Clin Invest 121:2133–2141

King DP, Takahashi JS (2000) Molecular genetics of circadian rhythms in mammals. Annu Rev Neurosci 23:713–742

Knutson SK, Chyla BJ, Amann JM, Bhaskara S, Huppert SS, Hiebert SW (2008) Liver-specific deletion of histone deacetylase 3 disrupts metabolic transcriptional networks. EMBO J 27:1017–1028

Kornmann B, Gachon F, Reinke H, Ripperger J, LeMartelot G, Schibler U (2006) The mammalian circadian timing system: from cyclic transcription to rhythmic physiology. FEBS J 273:6

Lam MTY, Cho H, Lesch HP, Gosselin D, Heinz S, Tanaka-Oishi Y, Benner C, Kaikkonen MU, Kim AS, Kosaka M, Lee CY, Watt A, Grossman TR, Rosenfeld MG, Evans RM, Glass CK (2013) Rev-Erbs repress macrophage gene expression by inhibiting enhancer-directed transcription. Nature 498:511–515

Lazar MA, Hodin RA, Darling DS, Chin WW (1989) A novel member of the thyroid steroid-hormone receptor family is encoded by the opposite strand of the rat c-erbA-alpha transcriptional unit. Mol Cell Biol 9:1128–1136

Le Martelot G, Claudel T, Gatfield D, Schaad O, Kornmann B, Lo Sasso G, Moschetta A, Schibler U (2009) REV-ERB alpha participates in circadian SREBP signaling and bile acid homeostasis. PLoS Biol 7:e1000181

Marvin KA, Reinking JL, Lee AJ, Pardee K, Krause HM, Burstyn JN (2009) Nuclear receptors homo sapiens Rev-erb beta and Drosophila melanogaster E75 are thiolate-ligated heme proteins which undergo Redox-mediated ligand switching and bind CO and NO. Biochemistry 48:7056–7071

Miyajima N, Horiuchi R, Shibuya Y, Fukushige S, Matsubara K, Toyoshima K, Yamamoto T (1989) 2 erbA homologs encoding proteins with different T3 binding-capacities are transcribed from opposite DNA strands of the same genetic-locus. Cell 57:31–39

Mullican SE, DiSpirito JR, Lazar MA (2013) The orphan nuclear receptors at their 25-year reunion. J Mol Endocrinol 51:T115–T140

Osborn O, Olefsky JM (2012) The cellular and signaling networks linking the immune system and metabolism in disease. Nat Med 18:363–374

Preitner N, Damiola F, Molina LL, Zakany J, Duboule D, Albrecht U, Schibler U (2002) The orphan nuclear receptor REV-ERB alpha controls circadian transcription within the positive limb of the mammalian circadian oscillator. Cell 110:251–260

Raghuram S, Stayrook KR, Huang PX, Rogers PM, Nosie AK, McClure DB, Burris LL, Khorasanizadeh S, Burris TP, Rastinejad F (2007) Identification of heme as the ligand for the orphan nuclear receptors REV-ERB alpha and REV-ERB beta. Nat Struct Mol Biol 14:1207–1213

Retnakaran R, Flock G, Giguere V (1994) Identification of RVR, a novel orphan nuclear receptor that acts as a negative transcriptional regulator. Mol Endocrinol 8:1234–1244

Rosbash M, Allada R, Dembinska M, Guo WQ, Le M, Marrus S, Qian Z, Rutila J, Yaglom J, Zeng H (1996) A Drosophila circadian clock. Cold Spring Harbor Symp Quant Biol 61:265–278

Schibler U (2006) Circadian time keeping: the daily ups and downs of genes, cells, and organisms. Prog Brain Res 153:271–282

Solt LA, Wang YJ, Banerjee S, Hughes T, Kojetin DJ, Lundasen T, Shin Y, Liu J, Cameron MD, Noel R, Yoo SH, Takahashi JS, Butler AA, Kamenecka TM, Burris TP (2012) Regulation of circadian behaviour and metabolism by synthetic REV-ERB agonists. Nature 485:62–68

Sun Z, Miller RA, Patel RT, Chen J, Dhir R, Wang H, Zhang DY, Graham MJ, Unterman TG, Shulman GI, Sztalryd C, Bennett MJ, Ahima RS, Birnbaum MJ, Lazar MA (2012) Hepatic Hdac3 promotes gluconeogenesis by repressing lipid synthesis and sequestration. Nat Med 18:934–942

Sun Z, Feng D, Fang B, Mullican SE, You SH, Lim HW, Everett LJ, Nabel CS, Li Y, Selvakumaran V, Won KJ, Lazar MA (2013) Deacetylase-independent function of HDAC3 in transcription and metabolism requires nuclear receptor corepressor. Mol Cell 52:769–782

Ukai-Tadenuma M, Yamada RG, Xu HY, Ripperger JA, Liu AC, Ueda HR (2011) Delay in feedback repression by cryptochrome 1 is required for circadian clock function. Cell 144:268–281

Vieira E, Marroqui L, Batista TM, Caballero-Garrido E, Carneiro EM, Boschero AC, Nadal A, Quesada I (2012) The clock gene Rev-erb alpha regulates pancreatic beta-cell function: modulation by leptin and high-fat diet. Endocrinology 153:592–601

Vieira E, Marroqui L, Figueroa ALC, Merino B, Fernandez-Ruiz R, Nadal A, Burris TP, Gomis R, Quesada I (2013) Involvement of the clock gene Rev-erb alpha in the regulation of glucagon secretion in pancreatic alpha-cells. PLoS One 8:e69939

Wang J, Lazar MA (2008) Bifunctional role of Rev-erb alpha in adipocyte differentiation. Mol Cell Biol 28:2213–2220

Woldt E, Sebti Y, Solt LA, Duhem C, Lancel S, Eeckhoute J, Hesselink MKC, Paquet C, Delhaye S, Shin YS, Kamenecka TM, Schaart G, Lefebvre P, Neviere R, Burris TP, Schrauwen P, Staels B, Duez H (2013) Rev-erb-alpha modulates skeletal muscle oxidative capacity by regulating mitochondrial biogenesis and autophagy. Nat Med 19:1039–1046

Yang XY, Downes M, Yu RT, Bookout AL, He WM, Straume M, Mangelsdorf DJ, Evans RM (2006) Nuclear receptor expression links the circadian clock to metabolism. Cell 126:801–810

Yin L, Lazar MA (2005) The orphan nuclear receptor Rev-erb alpha recruits the N-CoR/histone deacetylase 3 corepressor to regulate the circadian Bmal1 gene. Mol Endocrinol 19:1452–1459

Yin L, Wu N, Curtin JC, Qatanani M, Szwergold NR, Reid RA, Waitt GM, Parks DJ, Pearce KH, Wisely GB, Lazar MA (2007) Rev-erb alpha, a heme sensor that coordinates metabolic and circadian pathways. Science 318:1786–1789

Yoon HG, Chan DW, Huang ZQ, Li JW, Fondell JD, Qin J, Wong JM (2003) Purification and functional characterization of the human N-CoR complex: the roles of HDAC3, TBL1 and TBLR1. EMBO J 22:1336–1346

Zamir I, Harding HP, Atkins GB, Horlein A, Glass CK, Rosenfeld MG, Lazar MA (1996) A nuclear hormone receptor corepressor mediates transcriptional silencing by receptors with distinct repression domains. Mol Cell Biol 16:5458–5465

Zamir I, Zhang JS, Lazar MA (1997) Stoichiometric and steric principles governing repression by nuclear hormone receptors. Genes Dev 11:835–846

Zhang JS, Kalkum M, Chait BT, Roeder RG (2002) The N-CoR-HDAC3 nuclear receptor corepressor complex inhibits the JNK pathway through the integral subunit GPS2. Mol Cell 9:611–623

Zhao Q, Khorasanizadeh S, Miyoshi Y, Lazar MA, Rastinejad F (1998) Structural elements of an orphan nuclear receptor-DNA complex. Mol Cell 1:849–861

Control of Sleep-Wake Cycles in Drosophila

Abhishek Chatterjee and François Rouyer

Abstract Inter-oscillator communication modulates and sustains the circadian locomotor rhythms in flies and rodent animal models. In Drosophila, the multi-oscillator network that controls sleep-wake cycles includes about 150 clock neurons. A subset of lateral neurons (LNs) expressing the Pigment-dispersing factor (PDF) appears to act as a master clock in constant darkness (DD). In light–dark (LD) cycles, flies show a bimodal distribution of their activity, and the PDF-expressing LNs play a major role in the control of the morning bout of activity. In contrast, a subset of PDF-negative LNs can generate evening activity in the absence of other functional oscillators. How these oscillators interact in a fully functional network to shape the sleep-wake cycle remains debated. The PDF neurons strongly influence the PDF-negative ones in DD and, to a lesser extent, in LD. The extent of hierarchy depends on environmental conditions and the way the dominance of PDF neurons is exerted on the different types of PDF-negative neurons is unclear. The recent discovery of light- and temperature-dependent oscillators in the dorsal neurons (DNs) sheds new light on the circuits that control the Drosophila diurnal behavior and its adaptation to environmental changes.

Background

The fruit fly *Drosophila melanogaster* displays rest-activity rhythms that rely on a circadian clock located in the brain. In light–dark (LD) cycles, adult flies show a bimodal activity with morning and evening peaks at dawn and dusk. Activity rhythms persist in constant darkness (DD), indicating the circadian nature of this behavior. Like peripheral clocks, the brain clock depends on a molecular feedback loop where the CLOCK (CLK) and CYCLE (CYC) transcriptional factors drive the expression of the PERIOD (PER) and TIMELESS (TIM) proteins that repress CLK/CYC activity. The negative feedback loop operates in about 150 neurons,

A. Chatterjee • F. Rouyer (⌧)
Institut de Neurosciences Paris-Saclay, CNRS/Université Paris Sud, Gif-sur-Yvette, France
e-mail: Francois.rouyer@inaf.cnrs-gif.fr

© The Author(s) 2016
P. Sassone-Corsi, Y. Christen (eds.), *A Time for Metabolism and Hormones*,
Research and Perspectives in Endocrine Interactions,
DOI 10.1007/978-3-319-27069-2_8

the so-called clock neurons, which account for about 0.1 % of the total CNS neural population. The numerical simplicity of these 150 neurons that form a network is remarkable in comparison to central circuits for other hardwired behaviors such as courtship or learning and memory. The Drosophila clock neuronal network is also remarkably simple in comparison to circadian control circuits in vertebrates, where several brain areas, including the hypothalamus, pituitary gland, pineal gland, olfactory bulb, etc., harbor numerous bona fide clock neurons. This smaller number of neurons offers the potential to manipulate oscillators at the single-cell level in vivo, through well-defined genetic handles.

There are two broad populations within the 150 clock neurons of the fly brain; one population is laterally placed and another is located along the dorsal margin of the brain. The lateral neurons (LNs) lie near the interface of the central brain and the optic lobe and are organized into a ventral cluster that include small (s-LNvs) and large (l-LNvs) cells, a dorsal cluster (LNds) and a posterior cluster (LPNs). The dorsal neurons (DNs) are in turn subdivided into three clusters designated as DN1, DN2 and DN3 (Fig. 1). Such anatomical categorization frequently has neurochemical and functional bases; for example, the four most ventral s-LNvs express the Pigment-dispersing factor (PDF) neuropeptide and promote morning activity in LD. Based on strong functional data, mostly behavioral and some neurophysiological in nature, a wiring diagram of these differentiated clusters of brain clock neurons has begun to materialize over the past 10 years. In the following section we will summarize the logic of organization of this circuit.

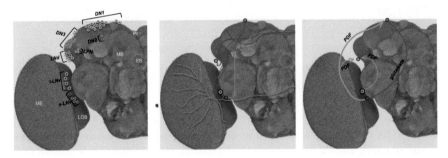

Fig. 1 The clock neurons of the Drosophila brain. *Left panel*: seven groups of clock neurons have been defined on an anatomical basis. The lateral neurons are organized into a ventral cluster that include small (s-LNvs) and large (l-LNvs) cells, a dorsal cluster (LNds) and a posterior cluster (LPNs). The dorsal neurons include three clusters designated as DN1, DN2 and DN3. Several neuropiles are indicated: Medulla (ME) and Lobula (LOB) in the optic lobe and Pars Intercerebralis (PI), Mushroom Bodies (MB) and Ellipsoid Body (EB) in the central brain. *Central panel*: projections of the different clock neuron subsets: s-LNvs and l-LNvs (*orange*), LNds and fifth PDF-negative s-LNv (*red*), DN1s (*blue*). *Right panel*: communication between neuronal clusters involves PDF from s-LNvs to LNds and DN1s as well as from l-LNvs to LNds, and glutamate from DN1s to s-LNvs

Layout of the Network in Constant Conditions

In the absence of cycling environmental cues, the fly clock circuit has been shown to adopt a functionally minimalist organization. In flies that experienced light–dark cycles and were subsequently kept in constant darkness, the presence of PDF-expressing s-LNvs was required to drive robust rhythmic behavior (Helfrich-Förster 1998; Renn et al. 1999) and a clock restricted to the PDF-expressing cells was sufficient to drive 24-h rhythms (Grima et al. 2004). In contrast, PDF-negative neurons drove behavioral rhythmicity under constant light if light inputs were reduced by the absence of the cell-resident photopigment cryptochrome (CRY). The precise location of these neurons that act as the pace-maker in constant light (LL) has been suggested to be either within the LNd cluster (Picot et al. 2007) or within the DN1s (Murad et al. 2007; Stoleru et al. 2007). In spite of running a functioning oscillator, the s-LNv neurons fail to influence the behavioral period in LL (Picot et al. 2007; Stoleru et al. 2007). The predominant contribution of the s-LNvs to behavioral rhythmicity, as evident in DD, becomes dramatically corroded in LL as ambient light inhibits their behavioral output (Picot et al. 2007). Under constant conditions the operation of the clock circuit remains highly centralized, but depending on the sustained presence or absence of light, this central position is occupied by either the PDF-negative clock neurons or the PDF-positive s-LNv neurons, respectively. Notably, in DD, the s-LNvs operate at the pinnacle of a hierarchy as they enforce a majority of other oscillators to realign their clock program in accordance with the sLNv pace (Stoleru et al. 2005). In contrast, messages from the non-PDF clocks have considerably subdued influence on the running of the master pacemaker in DD (Stoleru et al. 2005; Picot et al. 2007; Collins et al. 2014).

Recently, however, the existence of a centralized monopolar circuit organization in DD has been seriously challenged. The PDF clock has been shown to coherently change behavioral period only over a limited range, which is distributed asymmetrically around the 24-h focal point (Yao and Shafer 2014; Beckwith and Ceriani 2015). When the PDF neurons were forced to run at a pace beyond this specified range, multiple peaks of behavioral period emerged within a single fly, likely as a result of internal desynchronization among multiple oscillators (Yao and Shafer 2014; Beckwith and Ceriani 2015). Because these oscillators are coupled to the PDF clock with differing strength and range of entrainment, they are differentially affected by speed changes in the PDF clock (Yao and Shafer 2014). Thus, the behavioral period in DD is determined by the pace of not only the s-LNv clock but also by other oscillators enjoying different degrees of independence, although they were formerly thought to uniformly behave as slaves of the s-LNv pacemaker. As opposed to direct manipulation of individual oscillator pace, a parallel line of research was to putatively increase the excitability of different subsets of clock neurons to enhance their contribution in the network. This study raised the interesting possibility that the CRY-negative clock neurons, e.g., the DN2s, may have the potential to affect behavioral period like the well-known s-LNvs (Dissel et al. 2014). Going one step further, it was proposed that the DD behavioral period is constructed by integrating the slightly longer period dictated by the s-LNvs and the slightly shorter period imposed by the DN2s, with other clock neurons modulating the contributions of these two oscillators (Dissel et al. 2014). The behavioral period would thus depend on the interactions between differently paced oscillators whose endogenous period and influence in the network vary according to environmental conditions. However, speed changes in all clock neurons excluding the PDF cells fell short of altering the behavioral period (Yao and Shafer 2014), in fact bolstering the older idea that PDF oscillators are the predominant determinant of the behavior period in DD. In absence of PDF signaling, the output from the s-LNvs was compromised, thereby allowing secondary oscillators to strongly influence the behavior period (Yao and Shafer 2014). Precisely which oscillators are coupled, whether coupling is directional, how the coupling strength is determined and what are the relative weights of different oscillators to behavioral period according to environmental conditions are some of the questions that fly chronobiologists will probably resolve in next few years. We predict that the existing momentum on neuronal mechanisms of behavioral period determination will be extended to understand the other fundamental parameters of rhythm, such as phase and waveform.

Network Architecture Under LD Cycles

Depending on the constraints of physiological thermal limit and light availability, animals evolved few basic patterns of diel activity: diurnal, nocturnal, crepuscular or cathemeral (Bennie et al. 2014). In mammals, a given animal can stably and predictably switch back and forth between different patterns in context-dependent ways (Kas and Edgar 1999; Mrosovsky 2003). The choice of a temporal niche takes place downstream of the suprachiasmatic nucleus (SCN) clock and is strongly influenced by light inputs (Mrosovsky and Hattar 2005; Doyle et al. 2008). A comparable plasticity is observed in flies. For example, a typically crepuscular male fly will become nocturnal in the presence of a mate or during moonlit nights (Bachleitner et al. 2007; Fujii et al. 2007) and will be more diurnal when daylight is low (Schlichting et al. 2015). At first glance, the similar phasing of molecular oscillations in the different clock neuron subsets of the brain suggests that shaping the sleep-wake cycle occurs downstream of the clock, but results obtained from manipulating these different subsets support a more complex model.

Flies in the standard laboratory condition of 12:12 LD cycles show a bimodal profile with peaks of activity coinciding with the putative twilight transitions. Very nicely, the behavioral sub-routines of generating a morning peak and an evening peak are orchestrated by two separable subsets of oscillator neurons, the s-LNvs and the LNds, respectively, providing concrete experimental support for the dual-oscillator model of Daan and Pittendrigh (Pittendrigh and Daan 1976; Grima et al. 2004; Stoleru et al. 2004) (Fig. 2). Of note, this dual-oscillator ground plan

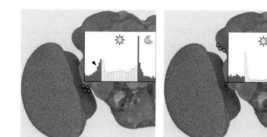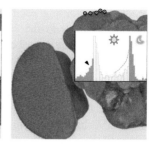

Fig. 2 Contribution of different clock neuron subsets to the LD behavior. Activity plots show the contribution of neuronal groups to morning and evening LD behavior. s-LNvs promote morning activity (*left*) whereas LNds and the fifth PDF-negative s-LNv promote evening activity (*center*). DN1s can promote morning activity and evening activity (*right*). The evening output of the DN1s is very weak in high light but strongly increases in low light (*dashed line*)

contrasts with the predominantly monopolar hierarchical organization prevailing in DD. The rather autonomous operation of two oscillators under LD cycles is abetted by their independent access to light information through redundant pathways—in-house CRY signaling and the visual system—whose output impinges on the clock circuit probably at multiple nodes, including the LNvs (Cusumano et al. 2009; Zhang et al. 2009). However, network interaction between oscillator neuron clusters could still be evident in LD. In the absence of PDF secretion by the LNvs, the evening peak of activity is advanced by a couple of hours (Renn et al. 1999). In the absence of both CRY and PDF, the evening peak vanishes and the phase of the molecular oscillations in the LNds is strongly altered (Cusumano et al. 2009; Zhang et al. 2009; Im et al. 2011). The phasing of the evening activity thus increasingly depends on PDF signaling when autonomous CRY-dependent photoreception decreases at the end of the day because of less intense and more reddish light. So far, the role of DNs in shaping the LD activity pattern seems to be largely secondary to the LNs (Grima et al. 2004; Stoleru et al. 2004; Zhang et al. 2010a, b). A functional clock restricted to the DN1s is sufficient to drive both morning and evening activity bouts in low light LD conditions, whereas high light permits morning activity only (Zhang et al. 2010b). These outputs are affected by temperature, and the DN1 neurons thus appear to be capable of integrating certain light and temperature information from the ambient environment (Zhang et al. 2010a, b). Although the expression of PDFR in the DN1s is important for their proper function (Zhang et al. 2010a), how they modulate the clock network's collective output remains unknown.

In summary, under periodic environmental cues, multiple, highly autonomous oscillators with distinct behavioral contributions collaboratively sculpt the organism's activity profile. In line with the 'internal coincidence model' of photoperiodism (Pittendrigh and Minis 1964), flexible changes in the clock network favoring the contribution of particular oscillators under certain ambient environmental conditions have been put forward as the mechanistic basis of seasonal adaptation in flies (Stoleru et al. 2007). The Daan/Pittendrigh model proposed that light accelerates morning oscillators and decelerates evening oscillators to adapt the bimodal activity to the changing photoperiod. Fast- and slow-running neuronal oscillators were described in flies displaying split rhythms in LL (Yoshii et al. 2004). However, short and long period components were observed to derive from the LD evening bout, suggesting that light-accelerated clock neurons contribute to the evening activity (Yoshii et al. 2004; Rieger et al. 2006), in contrast to the prediction of the model. As indicated above, light was shown to promote the output of the LNd-based evening oscillator while inhibiting the morning oscillator carried by the PDF-expressing s-LNvs (Picot et al. 2007). The importance of PDF-negative cells in the presence of light is also shown by experiments comparing the relative influence of PDF-positive and PDF-negative neuronal subsets in different photoperiods. This work was done by looking at morning and evening activity peaks of flies with accelerated PDF-positive or PDF-negative neurons. Under long photoperiods, the evening oscillator located in PDF-negative cells was proposed to control the speed of the morning oscillators, whereas in short photoperiod conditions the morning oscillator of PDF cells would take the lead (Stoleru et al. 2007). The discovery of other subsets contributing to morning and evening activity bouts, in particular the DN1s contributing to evening activity in low light only, suggest that the adaptation to photoperiod changes might be more complex. Indeed, we have data indicating that new groups of oscillators are recruited when flies are confronted with summer-like conditions. Such laboratory-based simplified environmental parameters are probably inadequate to explain the working of the network under the complex natural conditions that exist in the spatiotemporal niche inhabited by Drosophila in the wild (Menegazzi et al. 2012, 2013; Vanin et al. 2012; De et al. 2013). In particular, daily temperature variations have a strong impact on the sleep-activity pattern and can even induce some morning and evening anticipatory activity in clockless flies (Vanin et al. 2012; Menegazzi et al. 2013). But the principles and logic of circuit operation learned from a severely artificial set-up could well be valid and applicable for the same network's functionality under more complex natural conditions.

References

Bachleitner W, Kempinger L, Wulbeck C, Rieger D, Helfrich-Forster C (2007) Moonlight shifts the endogenous clock of *Drosophila melanogaster*. Proc Natl Acad Sci USA 104:3538–3543

Beckwith EJ, Ceriani MF (2015) Experimental assessment of the network properties of the Drosophila circadian clock. J Comp Neurol 523:982–996

Bennie JJ, Duffy JP, Inger R, Gaston KJ (2014) Biogeography of time partitioning in mammals. Proc Natl Acad Sci USA 111:13727–13732

Collins B, Kaplan HS, Cavey M, Lelito KR, Bahle AH, Zhu Z, Macara AM, Roman G, Shafer OT, Blau J (2014) Differentially timed extracellular signals synchronize pacemaker neuron clocks. PLoS Biol 12:e1001959

Cusumano P, Klarsfeld A, Chélot E, Picot M, Richier B, Rouyer F (2009) PDF-modulated visual inputs and cryptochrome define diurnal behavior in Drosophila. Nat Neurosci 12:1427–1433

De J, Varma V, Saha S, Sheeba V, Sharma VK (2013) Significance of activity peaks in fruit flies, *Drosophila melanogaster*, under seminatural conditions. Proc Natl Acad Sci USA 110: 8984–8989

Dissel S, Hansen CN, Ozkaya O, Hemsley M, Kyriacou CP, Rosato E (2014) The logic of circadian organization in Drosophila. Curr Biol 24:2257–2266

Doyle SE, Yoshikawa T, Hillson H, Menaker M (2008) Retinal pathways influence temporal niche. Proc Natl Acad Sci USA 105:13133–13138

Fujii S, Krishnan P, Hardin P, Amrein H (2007) Nocturnal male sex drive in Drosophila. Curr Biol 17:244–251

Grima B, Chélot E, Xia R, Rouyer F (2004) Morning and evening peaks of activity rely on different clock neurons of the Drosophila brain. Nature 431:869–873

Helfrich-Förster C (1998) Robust circadian rhythmicity of *Drosophila melanogaster* requires the presence of lateral neurons: a brain-behavioral study of disconnected mutants. J Comp Physiol A Neuroethol Sens Neural Behav Physiol 182:435–453

Im SH, Li W, Taghert PH (2011) PDFR and CRY signaling converge in a subset of clock neurons to modulate the amplitude and phase of circadian behavior in Drosophila. PLoS One 6:e18974

Kas MJ, Edgar DM (1999) A nonphotic stimulus inverts the diurnal-nocturnal phase preference in Octodon degus. J Neurosci 19:328–333

Menegazzi P, Yoshii T, Helfrich-Forster C (2012) Laboratory versus nature: the two sides of the Drosophila circadian clock. J Biol Rhythms 27:433–442

Menegazzi P, Vanin S, Yoshii T, Rieger D, Hermann C, Dusik V, Kyriacou CP, Helfrich-Forster C, Costa R (2013) Drosophila clock neurons under natural conditions. J Biol Rhythms 28:3–14

Mrosovsky N (2003) Beyond the suprachiasmatic nucleus. Chronobiol Int 20:1–8

Mrosovsky N, Hattar S (2005) Diurnal mice (Mus musculus) and other examples of temporal niche switching. J Comp Physiol A Neuroethol Sens Neural Behav Physiol 191:1011–1024

Murad A, Emery-Le M, Emery P (2007) A subset of dorsal neurons modulates circadian behavior and light responses in Drosophila. Neuron 53:689–701

Picot M, Cusumano P, Klarsfeld A, Ueda R, Rouyer F (2007) Light activates output from evening neurons and inhibits output from morning neurons in the Drosophila circadian clock. PLoS Biol 5:e315

Pittendrigh C, Daan S (1976) A functional analysis of circadian pacemakers in nocturnal rodents. V Pacemaker structure: a clock for all seasons. J Comp Physiol A Neuroethol Sens Neural Behav Physiol 106:333–335

Pittendrigh CS, Minis DH (1964) The entrainment of circadian oscillations by light and their role as photoperiodic clocks. Am Nat 98:261–294

Renn SC, Park JH, Rosbash M, Hall JC, Taghert PH (1999) A pdf neuropeptide gene mutation and ablation of PDF neurons each cause severe abnormalities of behavioral circadian rhythms in Drosophila. Cell 99:791–802

Rieger D, Shafer OT, Tomioka K, Helfrich-Forster C (2006) Functional analysis of circadian pacemaker neurons in *Drosophila melanogaster*. J Neurosci 26:2531–2543

Schlichting M, Grebler R, Menegazzi P, Helfrich-Forster C (2015) Twilight dominates over moonlight in adjusting Drosophila's activity pattern. J Biol Rhythms 30:117–128

Stoleru D, Peng P, Agosto J, Rosbash M (2004) Coupled oscillators control morning and evening locomotor behavior of Drosophila. Nature 431:862–868

Stoleru D, Peng Y, Nawathean P, Rosbash M (2005) A resetting signal between Drosophila pacemakers synchronizes morning and evening activity. Nature 438:238–242

Stoleru D, Nawathean P, Fernandez Mde L, Menet JS, Ceriani MF, Rosbash M (2007) The Drosophila circadian network is a seasonal timer. Cell 129:207–219

Vanin S, Bhutani S, Montelli S, Menegazzi P, Green EW, Pegoraro M, Sandrelli F, Costa R, Kyriacou CP (2012) Unexpected features of Drosophila circadian behavioural rhythms under natural conditions. Nature 484:371–375

Yao Z, Shafer OT (2014) The Drosophila circadian clock is a variably coupled network of multiple peptidergic units. Science 343:1516–1520

Yoshii T, Funada Y, Ibuki-Ishibashi T, Matsumoto A, Tanimura T, Tomioka K (2004) Drosophila cry(b) mutation reveals two circadian clocks that drive locomotor rhythm and have different responsiveness to light. J Insect Physiol 50:479–488

Zhang L, Lear BC, Seluzicki A, Allada R (2009) The CRYPTOCHROME photoreceptor gates PDF neuropeptide signaling to set circadian network hierarchy in Drosophila. Curr Biol 19: 2050–2055

Zhang L, Chung BY, Lear BC, Kilman VL, Liu Y, Mahesh G, Meissner RA, Hardin PE, Allada R (2010a) DN1(p) Circadian neurons coordinate acute light and PDF inputs to produce robust daily behavior in Drosophila. Curr Biol 20:591–599

Zhang Y, Liu Y, Bilodeau-Wentworth D, Hardin PE, Emery P (2010b) Light and temperature control the contribution of specific DN1 neurons to Drosophila circadian behavior. Curr Biol 20:600–605

Circadian Metabolomics: Insights for Biology and Medicine

Steven A. Brown and Ludmila Gaspar

Abstract A biological "circadian" clock governs nearly all aspects of mammalian behavior and physiology. This control extends from activities of entire organ systems down to individual cells, all of which contain autonomous molecular clocks. Under this control, a significant fraction of the cellular metabolome—the collection of all small-molecule metabolites—varies in abundance according to time of day. Comparing the rhythmic expression of transcripts, proteins, and metabolites has yielded valuable insights into clock-controlled physiological mechanisms. In the future, their analysis could provide a glimpse of instantaneous clock phase, even providing notions of clock time based upon molecules within a single breath. Such knowledge could be important for disease diagnosis and for chronopharmacology.

Introduction: A Many-Clock Problem

A "master clock" tissue in mammals has been identified in the suprachiasmatic nuclei (SCN) of the hypothalamus, about 20,000 neurons distributed into bilateral nuclei just above the optic chiasma. Lesioning of this region results in loss of circadian behavior and physiology under constant environmental conditions (Eastman et al. 1984), and transplantation results in circadian behavior corresponding to that of the donor animal (Ralph et al. 1990). Although the SCN directs circadian timing, the circuitry of which circadian clocks are composed is in fact much more widespread: nearly every cell in the body contains an autonomous molecular oscillator driven by feedback loops of transcription and translation of dedicated "core clock" proteins (Brown and Azzi 2013). Therefore, circadian control of complex physiology is at least in part a question of orchestration: on the one hand, circadian signals from the SCN must synchronize peripheral oscillators

S.A. Brown (✉) • L. Gaspar
Chronobiology and Sleep Research Group, Institute of Pharmacology and Toxicology, University of Zürich, Winterthurerstrasse 190, 8057 Zürich, Switzerland
e-mail: steven.brown@pharma.uzh.ch

© The Author(s) 2016   79
P. Sassone-Corsi, Y. Christen (eds.), *A Time for Metabolism and Hormones*,
Research and Perspectives in Endocrine Interactions,
DOI 10.1007/978-3-319-27069-2_9

elsewhere in the brain and body and, on the other, peripheral clocks must themselves direct circadian processes at a cellular level.

The ultimate consequences of this pervasive circadian control are that, in most mammalian tissues, 6–20 % of all transcripts and proteins are expressed in circadian fashion, i.e., with higher expression at one time of day and lower expression at another (Panda et al. 2002; Storch et al. 2002; Reddy et al. 2006; Robles et al. 2014). It is therefore not surprising that about 20 % of the mammalian metabolome shows circadian variation in both mice and men (Minami et al. 2009; Dallmann et al. 2012; Eckel-Mahan et al. 2012). Given the lower complexity of the metabolome and its extremely high conservation across species compared to the genome, an increasing number of studies have turned to metabolomics analyses to understand circadian biology.

Normally, circadian clocks throughout the body remain in relative synchrony with defined phase relationships. However, during timing shifts provoked by travel and shiftwork, in pathological cases such as inflammation and disease, or even due to the abnormal timing of food intake, this synchrony can be disrupted. For example, if normally nocturnal rodents are fed only during the day, clock phase in peripheral organs like liver and heart will change phase by nearly 12 h, while the SCN remains unaltered (Damiola et al. 2000; Stokkan et al. 2001). On the other hand, during a sudden change in light timing, the SCN will quickly alter its phase whereas peripheral organs require multiple days to do so (Davidson et al. 2009). Disease-mediated inflammation provides another example of peripheral clock dampening or dephasing: in response to infection, the circadian amplitude of transcription for multiple clock and clock-controlled genes decreases markedly (Cavadini et al. 2007). Finally, in both brain and peripheral tissues, sleep-related cellular signals can conflict with clock-related ones, leading to a dampening of circadian amplitude of clock-controlled genes (Maret et al. 2007; Moller-Levet et al. 2013; Archer et al. 2014). Both immediate and long-term consequences of such "clock desynchrony" are only beginning to be understood. For example, circadian amplitude in human subjects is directly correlated with survival time in some cancers (Innominato et al. 2012), and multiple studies in both humans and animals have linked shiftwork to increased disease and mortality (Viswanathan and Schernhammer 2009; Evans and Davidson 2013). As we discuss further below, metabolomics analyses could provide a powerful tool to study circadian phase and amplitude in both humans and animal models, potentially linking these parameters to human health in a wide variety of contexts.

An Overview of Circadian Metabolomics

Typically, comprehensive metabolomics analyses are conducted by flow injection mass spectrometry. Thus, in a single assay lasting a few seconds, thousands of peaks corresponding to individual metabolites can be detected. At the moment, a significant limiting factor for these studies is the identification of the metabolites corresponding to each peak. Most commercially accessible platforms can discretely identify a few hundred different substances, including lipids, amino acids, sugars, enzymatic cofactors, and peptides and hormones. In at least one study, these circadian metabolites have been compared to circadian transcripts in the same tissues in rodents, allowing a direct and comprehensive look at cellular pathways regulated in circadian fashion (Eckel-Mahan et al. 2012).

From this study, it was clear that the circadian clock exerts coordinated control over a large number of metabolic pathways, including those controlling the abundance of lipids, carbohydrates, and amino acids. Of course, given that food is itself consumed in time-of-day-dependent fashion, it would be formally possible that these variations could be indirect consequences of rhythmic activity, rather than direct clock control. In mice, for example, without rhythmic feeding only a small percentage of circadian transcripts continued to show diurnal oscillations (Vollmers et al. 2009). In humans, however, a very different picture has emerged. By analyzing metabolomics parameters from saliva and blood taken from humans kept in a "constant routine" of immobile reclined posture, hourly isocaloric meals, and sleep deprivation, Dallmann et al. (2012) could definitively rule out food-dependent control: 17 % of metabolites in both matrices were rhythmic even in the absence of rhythmic feeding, sleep, and activity (Fig. 1). These included lipids, carbohydrates, and amino acids, the same pathways that demonstrated metabolic control in mice (Eckel-Mahan et al. 2012). The same study also showed that the abundance of some metabolites increased or decreased monotonically with sleep deprivation, implying that sleep pressure and circadian influences might independently regulate diurnal metabolic physiology.

Metabolomics: Applications for Circadian Medicine

Because various circadian metabolites show peak abundance at different times of day, it is possible to use these relative quantities as indicators of timing. The idea is analogous to the "chronological garden" of the Swedish botanist Carl Linnaeus, who used plants flowering at different times of day to determine geological time at any moment. In precisely the same fashion, Minami et al. (2009) used blood metabolites from mice as a way of detecting circadian body time, and Martinez-Lozano Sinues et al. (2014) used metabolites within human breath. While potentially quite powerful, these molecular timetable-based methods are hampered by the high inter-individual variability of metabolite abundance among different subjects, making single-time-point analyses relatively imprecise. So far, an accuracy of about 2 h in circadian time is the best that has been attained. As more individuals are metabolically characterized into different endophenotypic subtypes, it is likely that this accuracy will increase substantially.

Major applications of such technology would be twofold. First and most simply, it would be possible to determine human body time prior to clinical intervention. For most drugs, both pharmacokinetics and pharmacodynamics vary in circadian

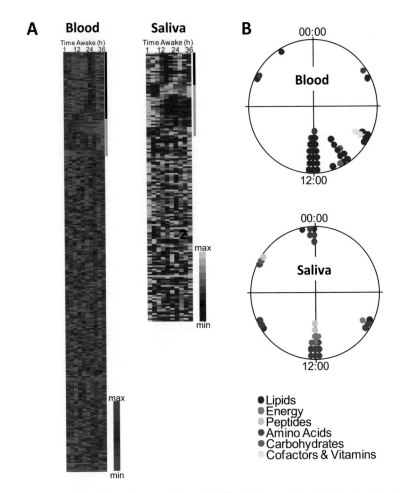

Fig. 1 (**a**) Heat map of circadian metabolites identified in human blood (*left*) and saliva (*right*) from subjects maintained in a constant routine of hourly isocaloric meals, immobile posture, constant dim light, and sleep deprivation. Rows: individual metabolites; columns: time relative to start of experiment. (**b**) Major classes of compounds identified in blood (*top*) and saliva (*bottom*), plotted in circadian time relative to theoretical dawn (Adapted from Dallmann et al. 2012)

fashion. In other words, not only is the metabolism of many xenobiotic substances strongly regulated but also the biological targets of the drugs themselves (Dallmann et al. 2014). Thus, potentially both increased efficacy and reduced toxicity could be obtained by precise timing of delivery, at least in the case of unstable compounds. Currently, multiple clinical trials have been run or are running, especially in the field of cancer, to test this concept (Innominato et al. 2014). Since cell division has been shown to be coordinated with circadian clock timing in both adult animals and cells (Matsuo et al. 2003; Nagoshi et al. 2004; Kowalska et al. 2013; Bieler et al. 2014; Feillet et al. 2014), and many chemotherapeutic agents are metabolized by circadian isoforms of cytochrome P450 in the liver, it is logical to imagine that chemotherapeutic toxicity would itself be circadian. This concept has been demonstrated experimentally in mice (Gorbacheva et al. 2005), and human trials have also shown time-of-day-dependent effects (Innominato et al. 2014).

Secondly, metabolite timetable-based methods would be able to predict not only clock phase but also clock amplitude and possibly even circadian desynchrony. These parameters have been increasingly linked to disease both in humans and in animal models, as mentioned in the introduction, leading the World Health Organization to classify shiftwork as a suspected carcinogen. Even a simple indicator like the amplitude of circadian behavior correlates directly with survival during chemotherapy of metastatic colon carcinoma in humans (Innominato et al. 2012).

It is suspected that one of the main deleterious effects of shiftwork is circadian desynchrony among different organs. In mice, changes in daylight timing shift different organs at different speeds (Davidson et al. 2009). Potentially, metabolomics could give insight into this phenomenon, since various circadian components come from different tissues. For example, the hormone melatonin is secreted by the pineal gland of the hypothalamus and is thought to be a direct output of the SCN. By contrast, many other endocrine factors and metabolites detectable in blood or in breath arise primarily as byproducts of peripheral organ function (Gamble et al. 2014). Therefore, we propose that circadian metabolomics could be useful in elucidating desynchrony between clocks in brain and in other peripheral tissues. To date, no studies have used metabolomics methods in this fashion, but great potential exists.

Outlook and Conclusion

Questions of circadian desynchrony in health and disease are only beginning to be addressed. The idea that shiftwork might affect circadian clock function is easy to see. However, many other factors could play important and unsuspected roles. For example, recent studies have suggested that chronic sleep restriction alone, even without changed diurnal patterns of activity, could also disrupt circadian transcription (Moller-Levet et al. 2013). Similarly, depressive and affective disorders have long been known to be accompanied by dramatically different sleep–wake patterns (Lamont et al. 2007). Recent research has established a close tie between circadian dysfunction and metabolic disorders like obesity and diabetes (Maury et al. 2014). All of these syndromes are potentially explorable by circadian metabolomics. The conclusions that such studies might derive could both answer outstanding questions about circadian biology and improve human health.

Acknowledgments The research of SAB is supported by the Swiss National Science Foundation, the Velux Foundation, the Swiss Cancer League, and the Zürich Clinical Research Priority Project "Sleep and Health." He is a member of the Zürich Neurozentrum (ZNZ) and Molecular Life Sciences graduate programs within the Life Sciences Zürich Graduate School. LC is a member of the ZNZ and has received support from the Zürich Clinical Research Priority Project "Sleep and Health".

References

Archer SN, Laing EE, Moller-Levet CS, van der Veen DR, Bucca G, Lazar AS, Santhi N, Slak A, Kabiljo R, von Schantz M, Smith CP, Dijk DJ (2014) Mistimed sleep disrupts circadian regulation of the human transcriptome. Proc Natl Acad Sci U S A 111:E682–E691

Bieler J, Cannavo R, Gustafson K, Gobet C, Gatfield D, Naef F (2014) Robust synchronization of coupled circadian and cell cycle oscillators in single mammalian cells. Mol Syst Biol 10:739

Brown SA, Azzi A (2013) Peripheral circadian oscillators in mammals. Handb Exp Pharmacol 217:45–66

Cavadini G, Petrzilka S, Kohler P, Jud C, Tobler I, Birchler T, Fontana A (2007) TNF-alpha suppresses the expression of clock genes by interfering with E-box-mediated transcription. Proc Natl Acad Sci U S A 104:12843–12848

Dallmann R, Viola AU, Tarokh L, Cajochen C, Brown SA (2012) The human circadian metabolome. Proc Natl Acad Sci U S A 109:2625–2629

Dallmann R, Brown SA, Gachon F (2014) Chronopharmacology: new insights and therapeutic implications. Annu Rev Pharmacol Toxicol 54:339–361

Damiola F, Le Minh N, Preitner N, Kornmann B, Fleury-Olela F, Schibler U (2000) Restricted feeding uncouples circadian oscillators in peripheral tissues from the central pacemaker in the suprachiasmatic nucleus. Genes Dev 14:2950–2961

Davidson AJ, Castanon-Cervantes O, Leise TL, Molyneux PC, Harrington ME (2009) Visualizing jet lag in the mouse suprachiasmatic nucleus and peripheral circadian timing system. Eur J Neurosci 29:171–180

Eastman CI, Mistlberger RE, Rechtschaffen A (1984) Suprachiasmatic nuclei lesions eliminate circadian temperature and sleep rhythms in the rat. Physiol Behav 32:357–368

Eckel-Mahan KL, Patel VR, Mohney RP, Vignola KS, Baldi P, Sassone-Corsi P (2012) Coordination of the transcriptome and metabolome by the circadian clock. Proc Natl Acad Sci U S A 109:5541–5546

Evans JA, Davidson AJ (2013) Health consequences of circadian disruption in humans and animal models. Prog Mol Biol Transl Sci 119:283–323

Feillet C, Krusche P, Tamanini F, Janssens RC, Downey MJ, Martin P, Teboul M, Saito S, Levi FA, Bretschneider T, van der Horst GT, Delaunay F, Rand DA (2014) Phase locking and multiple oscillating attractors for the coupled mammalian clock and cell cycle. Proc Natl Acad Sci U S A 111:9828–9833

Gamble KL, Berry R, Frank SJ, Young ME (2014) Circadian clock control of endocrine factors. Nat Rev Endocrinol 10:466–475

Gorbacheva VY, Kondratov RV, Zhang R, Cherukuri S, Gudkov AV, Takahashi JS, Antoch MP (2005) Circadian sensitivity to the chemotherapeutic agent cyclophosphamide depends on the functional status of the CLOCK/BMAL1 transactivation complex. Proc Natl Acad Sci U S A 102:3407–3412

Innominato PF, Giacchetti S, Bjarnason GA, Focan C, Garufi C, Coudert B, Iacobelli S, Tampellini M, Durando X, Mormont MC, Waterhouse J, Levi FA (2012) Prediction of overall survival through circadian rest-activity monitoring during chemotherapy for metastatic colorectal cancer. Int J Cancer 131:2684–2692

Innominato PF, Roche VP, Palesh OG, Ulusakarya A, Spiegel D, Levi FA (2014) The circadian timing system in clinical oncology. Ann Med 46:191–207

Kowalska E, Ripperger JA, Hoegger DC, Bruegger P, Buch T, Birchler T, Mueller A, Albrecht U, Contaldo C, Brown SA (2013) NONO couples the circadian clock to the cell cycle. Proc Natl Acad Sci U S A 110:1592–1599

Lamont EW, Legault-Coutu D, Cermakian N, Boivin DB (2007) The role of circadian clock genes in mental disorders. Dialogues Clin Neurosci 9:333–342

Maret S, Dorsaz S, Gurcel L, Pradervand S, Petit B, Pfister C, Hagenbuchle O, O'Hara BF, Franken P, Tafti M (2007) Homer1a is a core brain molecular correlate of sleep loss. Proc Natl Acad Sci U S A 104:20090–20095

Martinez-Lozano Sinues P, Tarokh L, Li X, Kohler M, Brown SA, Zenobi R, Dallmann R (2014) Circadian variation of the human metabolome captured by real-time breath analysis. PLoS One 9:e114422

Matsuo T, Yamaguchi S, Mitsui S, Emi A, Shimoda F, Okamura H (2003) Control mechanism of the circadian clock for timing of cell division in vivo. Science 302:255–259

Maury E, Hong HK, Bass J (2014) Circadian disruption in the pathogenesis of metabolic syndrome. Diabetes Metab 40:338–346

Minami Y, Kasukawa T, Kakazu Y, Iigo M, Sugimoto M, Ikeda S, Yasui A, van der Horst GT, Soga T, Ueda HR (2009) Measurement of internal body time by blood metabolomics. Proc Natl Acad Sci U S A 106:9890–9895

Moller-Levet CS, Archer SN, Bucca G, Laing EE, Slak A, Kabiljo R, Lo JC, Santhi N, von Schantz M, Smith CP, Dijk DJ (2013) Effects of insufficient sleep on circadian rhythmicity and expression amplitude of the human blood transcriptome. Proc Natl Acad Sci U S A 110: E1132–E1141

Nagoshi E, Saini C, Bauer C, Laroche T, Naef F, Schibler U (2004) Circadian gene expression in individual fibroblasts: cell-autonomous and self-sustained oscillators pass time to daughter cells. Cell 119:693–705

Panda S, Antoch MP, Miller BH, Su AI, Schook AB, Straume M, Schultz PG, Kay SA, Takahashi JS, Hogenesch JB (2002) Coordinated transcription of key pathways in the mouse by the circadian clock. Cell 109:307–320

Ralph MR, Foster RG, Davis FC, Menaker M (1990) Transplanted suprachiasmatic nucleus determines circadian period. Science 247:975–978

Reddy AB, Karp NA, Maywood ES, Sage EA, Deery M, O'Neill JS, Wong GK, Chesham J, Odell M, Lilley KS, Kyriacou CP, Hastings MH (2006) Circadian orchestration of the hepatic proteome. Curr Biol 16:1107–1115

Robles MS, Cox J, Mann M (2014) In-vivo quantitative proteomics reveals a key contribution of post-transcriptional mechanisms to the circadian regulation of liver metabolism. PLoS Genet 10:e1004047

Stokkan KA, Yamazaki S, Tei H, Sakaki Y, Menaker M (2001) Entrainment of the circadian clock in the liver by feeding. Science 291:490–493

Storch KF, Lipan O, Leykin I, Viswanathan N, Davis FC, Wong WH, Weitz CJ (2002) Extensive and divergent circadian gene expression in liver and heart. Nature 417:78–83

Viswanathan AN, Schernhammer ES (2009) Circulating melatonin and the risk of breast and endometrial cancer in women. Cancer Lett 281:1–7

Vollmers C, Gill S, DiTacchio L, Pulivarthy SR, Le HD, Panda S (2009) Time of feeding and the intrinsic circadian clock drive rhythms in hepatic gene expression. Proc Natl Acad Sci U S A 106:21453–21458

Rhythms Within Rhythms: The Importance of Oscillations for Glucocorticoid Hormones

Stafford Lightman

Abstract The circadian activity of the hypothalamic-pituitary-adrenal (HPA) axis is made up from an underlying oscillatory rhythm of ACTH and glucocorticoid pulses that vary in amplitude but not frequency over the 24 h. This oscillatory activity is not the result of a hypothalamic oscillator but emerges as a natural consequence of the feedforward:feedback interaction between the pituitary corticotropes and the glucocorticoid-secreting cells of the adrenal cortex. This oscillatory activity has resulted in adaptations in the way tissues read their 'digital' ligand signal. The adrenal cortex is relatively insensitive to constant signals of ACTH but responds briskly to the equivalent amount of ACTH administered in a pulsatile manner. Similarly glucocorticoid-responsive tissues such as the brain and the liver are able to read the oscillating signals of cortisol or corticosterone secretion, with differential biochemical and functional responses to different patterns of ligand presentation. During a prolonged acute stress there is a major change in the pituitary-adrenal relationship, with a marked increase in the sensitivity of the adrenal to small changes in ACTH, so that following cardiac surgery small oscillations in ACTH result in massive swings in cortisol. This response appears to be due to a change both in the ACTH signalling pathway and in the endogenous activators and inhibitors of glucocorticoid synthesis.

Introduction

Oscillations are a basic characteristic of all matter. Atoms have their own characteristic oscillation frequencies, and the frequency of the oscillations of Cesium 133, for instance, is often chosen as the basis for atomic clocks. The kinetic theory of matter goes further to suggest that all matter is made up of particles that are constantly moving; in 1905, it was Albert Einstein who demonstrated how this

S. Lightman (✉)
Henry Wellcome Laboratories for Integrative Neuroscience and Endocrinology, University of Bristol, Dorothy Hodgkin Building, Whitson Street, Bristol BS1 3NY, UK
e-mail: stafford.lightman@bristol.ac.uk

© The Author(s) 2016
P. Sassone-Corsi, Y. Christen (eds.), *A Time for Metabolism and Hormones*,
Research and Perspectives in Endocrine Interactions,
DOI 10.1007/978-3-319-27069-2_10

atomic activity explained the phenomenon of Brownian movement (Einstein 1905). It is therefore of no surprise that biological systems are also invariably dynamic, with both stochastic interactions and deterministic processes across multiple time scales ensuring the maintenance of homeostatic regulation and allowing the organism to adapt to changes in both internal and external environments.

The physical world has a direct impact on the one neuroendocrine system that is critical for life: the hypothalamic-pituitary-adrenal (HPA) axis. The daily rotation of the earth on its axis provides our planet with its regular 24-h day/night cycle and this is the cue for the circadian activity of the HPA axis, which ensures energy supplies are available prior to the daily phase of activity—day in man and night in rodents—by ensuring an anticipatory increase in plasma glucocorticoid levels. These glucocorticoid hormones—cortisol in man and corticosterone in the rodent (both called CORT in this manuscript)—do not simply organise the circadian aspects of metabolic, cognitive and immunological functions, they are also vital homeostatic regulators that are extremely responsive to any threat to the organism's internal stability. In addition to their circadian variation, they need to maintain exquisite sensitivity to both perceived and genuine stressors. It is this combined function of providing a day-to-day regulatory role together with a rapid response mode that requires a system that maintains its reactivity at all times, whatever the status of its circadian activity.

How can this be achieved? The circuitry for the HPA axis is shown in Fig. 1. As has been well described by other authors in this symposium, the suprachiasmatic nucleus (SCN) of the hypothalamus provides the circadian regulation via an inhibitory input to the corticotrophin-releasing hormone (CRH)-containing neurons of the hypothalamic paraventricular nucleus (PVN) (Vrang et al. 1995; Dickmeis 2009). These neurons in turn release CRH, which travels in the hypothalamic-pituitary portal blood system to corticotroph cells in the anterior pituitary, which then release adrenocorticotropic hormone (ACTH) into the systemic circulation. Surprisingly, the output from this system is not a simple analogue release of ACTH from the pituitary gland but a complex episodic series of pulses of hormone secretion (Jasper and Engeland 1991; Windle et al. 1998). In this chapter I shall describe the mechanism underlying the genesis of this oscillating hormone system and why it is so important for the ability of glucocorticoids to perform their multiple activities in so many different systems in the body.

The Origin of HPA Pulsatility

It had always been assumed that the pulsatility of both ACTH and CORT must be due to some hypothalamic oscillator resulting in pulses of CRH, which are then transcribed into pulses of ACTH and CORT. Indeed, there is evidence for episodic release of CRH from macaque hypothalamic explants (Mershon et al. 1992) and for rapid changes in CRH in the median eminence of rats (Ixart et al. 1991) and portal blood of sheep (Caraty et al. 1988). There is, however, a mismatch between the higher frequency of CRH pulses than the ACTH/CORT pulses, and Engler

Fig. 1 The hypothalamic-pituitary-adrenal (HPA) axis. The hypothalamus receives circadian input from the suprachiasmatic nucleus (SCN) and stress-related inputs from the limbic system and brainstem. *PVN* paraventricular nucleus, *CRH* corticotrophin-releasing hormone, *AVP* arginine vasopressin. Reproduced with permission from Lightman and Conway-Campbell (2010)

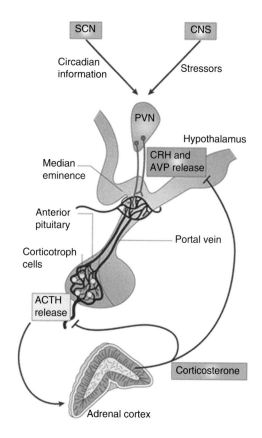

et al. (1990) also demonstrated the maintenance of ACTH and cortisol pulses in the sheep even after disconnection of the pituitary from hypothalamic portal blood.

We therefore reconsidered what we knew about the interaction between pituitary corticotropes and adrenal cortical cells. Since the adrenal gland cannot store glucocorticoids—which are very lipophilic and thus cannot be stored in vesicles—every pulse of steroid released into the circulation must be newly synthesised. There must, therefore, be a delay between the signal from ACTH activation of adrenal MC2 receptors and the release of CORT. Indeed, this has been clearly demonstrated both in the rat and in man (Carnes et al. 1989; Henley et al. 2009). There is, in other words, a clear delay in the feedforward effect of ACTH on the release of CORT. What about the feedback of CORT on the pituitary corticotropes? A very rapid effect of CORT inhibiting ACTH release has been demonstrated both in the rat and in man (Jones et al. 1972; Rotsztejn et al. 1975; Hinz and Hirschelmann 2000; Russell et al. 2010). We therefore have a feedforward/feedback interaction between pituitary corticotroph-derived ACTH and adrenal cortical-derived CORT, with a built-in delay in the feedforward part of the loop. This is a system that mathematically must show endogenous oscillatory activity!

With this knowledge, we were able to collaborate with our mathematical colleagues who developed a mathematical model that predicted the ability of the
pituitary adrenal system to support self-sustained ACTH and CORT oscillations
at the frequency found in normal physiology, even in the face of a constant CRH
drive (Walker et al. 2010). This model also predicted that these oscillations would
not occur at very low CRH concentrations and would be damped at the high levels
of CRH found after an acute stress, as we had found in rats following an acute
stressor (Windle et al. 1998). We were then able to test this model experimentally
using constant infusions of CRH into free running animals in the morning, a time
when their endogenous CRH systems are suppressed by the SCN. Consistent with
our mathematical model, a constant infusion of CRH produced normal ultradian
oscillations of both ACTH and CORT, with the same pulse frequency as that found
during normal nocturnal HPA activity (Walker et al. 2012). Furthermore, giving a
constant infusion of a higher concentration of CRH resulted in a high and prolonged
constant secretion of CORT, similar to that found following a severe stress.

Since circadian rhythmicity of the HPA axis is controlled by an inhibitory output
from the SCN to the PVN, another prediction from our mathematical model was
that interruption of this pathway would not only abolish circadian rhythmicity but
would also allow unrestrained CRH secretion throughout the 24 h, resulting in
ultradian secretion of ACTH and CORT throughout the 24 h and not just during
times of peak activity. We tested this prediction both by lesioning of the SCN and
by maintaining animals on a 24-h constant light cycle. Indeed, we found that in both
cases there was a loss of circadian variability but maintenance of ultradian activity
across the 24 h (Waite et al. 2012).

Implications of HPA Pulsatility

Since HPA pulsatility emerges as a natural consequence of the feedforward:feed-
back interaction between the pituitary and adrenal gland, it is not surprising that it
has been reported to exist in all mammalian species that have been studied,
including rat (Jasper and Engeland 1991; Windle et al. 1998), sheep (Fulkerson
1978), rhesus monkey (Holaday et al. 1977) and man (Weitzman et al. 1971;
Henley et al. 2009). It would seem very likely, therefore, that physiological systems
have adapted to read this digital ACTH and CORT signalling, and indeed this has
proved to be the case.

Adrenal Adaptation to Pulsatile ACTH

Adrenal steroidogenesis is an extraordinarily dynamic process. Since steroid hormones cannot be stored for subsequent rapid release, each pulse of CORT seen in
the plasma is the result of the very rapid intra-adrenal conversion of cholesterol to

CORT. Pituitary-derived ACTH binds to the melanocortin-2 receptor in adrenal fasciculata cells (Mountjoy et al. 1994), activating adenylyl cyclase and PKA-induced genomic and non-genomic steroidogenic pathways. CREB-induced transcription of the rate-limiting step of cholesterol transport into the mitochondrion (StAR) is enhanced by the binding of positive regulators (Sugawara et al. 1996; Caron et al. 1997; Song et al. 2001; Conkright et al. 2003; Takemori et al. 2007) and inhibition of the negative regulator DAX-1 (Song et al. 2004). PKA also modifies the rapid non-transcriptional modification of steroidogenic proteins, including phosphorylation of StAR itself (Arakane et al. 1997) and of hormone-sensitive lipase (HSL), which increases the intracellular levels of cholesterol itself.

So how does this complex system of different activators and inhibitors of CORT synthesis respond to different patterns of ACTH presentation? We have shown that, when endogenous ACTH is suppressed by administration of methylprednisolone, rats respond to pulsatile exogenous ACTH with a pulsatile release of CORT (Spiga et al. 2011). When the same dose of ACTH is infused at a constant rate, however, no CORT is secreted. Indeed, constant ACTH infusion actually results in a suppressed response to a subsequent stress amplitude pulse of ACTH, suggesting a dysregulation of the normal steroidogenic mechanisms (Spiga and Lightman 2014). The exact mechanism underlying this is unclear, but there is evidence that intra-adrenal GR can mediate local negative feedback on steroidogenesis via induction of DAX-1, perhaps accentuating the pulsatile characteristics of the response to a physiological pulse of ACTH while effectively inhibiting the response to a more constant exposure. Interestingly, by integrating our in vivo data with mathematical modelling of adrenal responses, we do find that rapid intra-adrenal inhibition must be an important factor in adrenal ultradian oscillations (Walker et al. 2014). This all suggests that the adrenal gland is beautifully adapted to respond to a pulsatile signal of ACTH, rather than showing a simple analogue response to different concentrations of this hormone.

During severe stress, there also seem to be special adaptations at the adrenal level. In a study of patients undergoing coronary artery bypass graft procedures (Fig. 2), we found an initial surge in both ACTH and CORT, followed by a fall in ACTH back to baseline levels but maintenance of the high levels of CORT, with continued but amplified ultradian responses of CORT to small changes in basal ACTH (Gibbison et al. 2014). The initial rise in both hormones was delayed after the actual surgery itself, suggesting it was the result of inflammatory cytokine production (from the sternotomy), which is known to go up at this time (Lahat et al. 1992; Roth-Isigkeit et al. 1999; de Mendonca-Filho et al. 2006). Therefore, in a reverse translation approach, we used a model of severe stress both with (LPS) and without (depot ACTH) associated systemic inflammation. ACTH and CORT followed each other closely in the depot ACTH-induced response, but after LPS we had the same findings as after cardiac surgery: maintenance of high CORT even after ACTH had fallen to normal levels. Furthermore, only in this group was there an increased expression of StAR and MRAP (a vital accessory for the MC2 receptor) mRNAs and StAR protein. This presumably explains the increased sensitivity to ACTH and the increased steroidogenesis at this time, which is quite

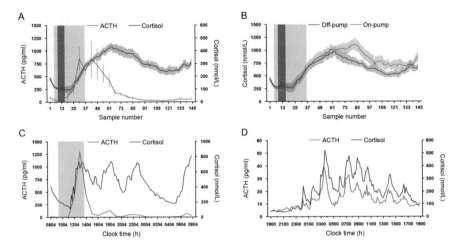

Fig. 2 Changes in cortisol and ACTH levels throughout the 24-h perioperative period of cardiac surgery. (**a**) Group mean ± SEM cortisol and ACTH. All sampling (i.e., the first sample in every case) started between 0800 and 0900 h. (**b**) Mean ± SEM 24-h cortisol profile from patients undergoing coronary artery bypass graft using the off-pump or the on-pump technique. All off-pump surgeries were performed between sample 5 and sample 35; all on-pump surgeries were performed between samples 5 and samples 36. (**a**) and (**b**), *light grey* area, period during which some patients were undergoing surgery. *Dark grey* area, period during which all patients were undergoing surgery. (**c**) Individual 24-h ACTH and cortisol profile of a patient undergoing off-pump CABG. *Light grey* area, period during which the patient was undergoing surgery (0919–1349-h). (**d**), Individual 24-h ACTH and cortisol profile of a healthy volunteer. Reproduced with permission from Gibbison et al. (2014)

different from the situation in patients who have had prolonged critical illness in whom steroidogenic genes appear to be depleted (Boonen et al. 2014).

Tissue Adaptation to Pulsatile CORT

The large oscillations of total CORT seen in blood are also reflected in similar large oscillations of the active free cortisol levels (unbound to cortisol binding globulin) in the brain and subcutaneous tissue (Droste et al. 2008; Qian et al. 2012; Bhake et al. 2013), indicating that both glucocorticoid (GR) and mineralocorticoid (MR) receptors will be exposed to oscillating levels of their ligand (Fig. 3). These receptors are latent transcription factors initially sequestered in the cytoplasm bound to chaperone molecules including HSP90 and p23. Upon binding of CORT, they undergo a conformational change dissociate from the chaperone complex and are actively transported into the nucleus where they rapidly cycle on and off glucocorticoid response elements (GREs) at the chromatin template (Fig. 4; Hager et al. 2006; Conway-Campbell et al. 2012). Each endogenous pulse of CORT results in a rapid increase in activated GR available for binding to GREs, with

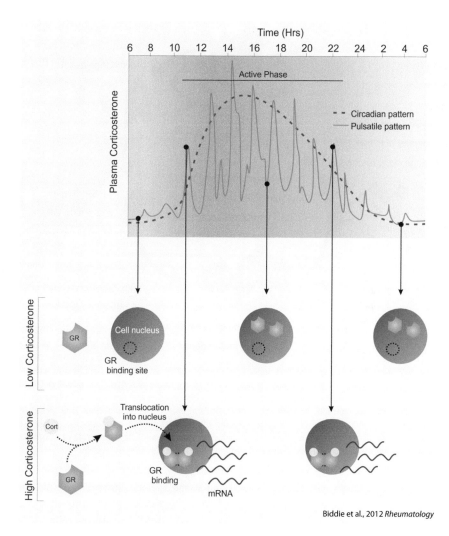

Biddie et al., 2012 *Rheumatology*

Fig. 3 Glucocorticoid pulsatility drives transient activation of GR-responsive genes. CORT levels rise in anticipation of the active phase. Hormone levels follow a circadian pattern, although the underlying pattern of hormone secretion is ultradian, where glucocorticoids are released approximately every hour. During a pulse, exposure to hormone drives GR translocation into the nucleus, where it binds to genomic elements to drive transcription. Hormone troughs result in GR dissociation from chromatin, releasing the receptor into the nucleoplasm ready to initiate transcription during further rises in hormone levels. The dynamics of the receptor and hormone secretion patterns allow rapid response to rapidly changing cellular and physiological conditions. Reproduced with permission from Biddie et al. (2011)

repeated pulses resulting in cyclical changes in GR chromatin association profiles on regulatory elements of endogenous CORT-regulated gene promoters (Fig. 4; Conway-Campbell et al. 2011). The interaction of GR with other accessory

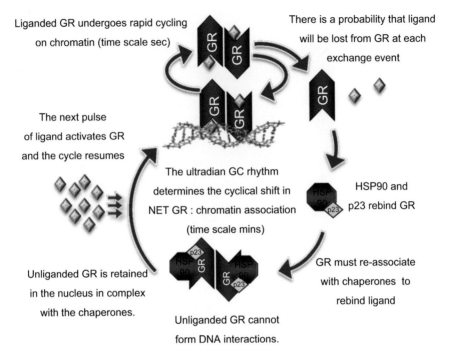

Liganded GR undergoes rapid cycling on chromatin (time scale sec)

There is a probability that ligand will be lost from GR at each exchange event

The next pulse of ligand activates GR and the cycle resumes

The ultradian GC rhythm determines the cyclical shift in NET GR : chromatin association (time scale mins)

HSP90 and p23 rebind GR

Unliganded GR is retained in the nucleus in complex with the chaperones.

GR must re-associate with chaperones to rebind ligand

Unliganded GR cannot form DNA interactions.

Fig. 4 GR ultradian cycling model. A schematic representation of how cyclical GR interactions with genomic response elements result in pulsatile transcriptional activity. The initial pulse of hormone causes nuclear translocation of GR, allowing GR to enter the chromatin binding cycle with rapid transient interactions with the chromatin template (stochastic action; time scale of seconds). The slower cyclical events related to the ultradian rhythm are not stochastic but are determined by pulses of ligand and the nuclear molecular chaperone cycle (time scale of minutes). Reproduced with permission from Conway-Campbell et al. (2012)

DNA-binding factors will clearly be cell and tissue specific, providing scope for differential responses to the same pattern of CORT pulses in different tissues.

There are now increasing data that ultradian pulsatility has considerable relevance for gene transcription. Ultradian oscillations of CORT induce cyclic GR-mediated pulses of gene transcription, both in vitro and in vivo, which differ from the response to equivalent constant levels of the ligand (Stavreva et al. 2009; Conway-Campbell et al. 2012; McMaster et al. 2011). Indeed, gene pulsing of the clock gene period 1 occurs in vivo in response to physiological pulses, both in the liver (Stavreva et al. 2009) and in the hippocampus (Conway-Campbell et al. 2010). In addition to these genomic effects, glucocorticoids have rapid non-genomic effects on neuronal activity in the brain (Karst et al. 2005; Evanson et al. 2010; Hill and Tasker 2012), with rapid effects on both excitatory and inhibitory inputs to the hippocampus (Karst et al. 2005), and evidence for a specific effect on the insertion of Ca^{2+}-permeable AMPA receptors into synapses (Whitehead et al. 2013). Glucocorticoid pulsatility has also been shown to have specific non-genomic effects on miniature excitatory postsynaptic current (mEPSC) frequency in different brain areas, with differential effects being described for the amygdala and the hippocampus (Karst et al. 2010). Recently, using high-resolution imaging and electrophysiology, this group showed that, while a single pulse of CORT increases hippocampal AMPAR signalling and impairs the induction of LTP for several hours, a second pulse restores the potentiation range of the glutamate synapses (Sarabdjitsingh et al. 2014). This finding suggests that pulsatile exposure to CORT is necessary to maintain optimal glutamatergic neurotransmission.

It is also becoming clear that the pattern of glucocorticoid secretion has a considerable impact on behaviour. Adrenalectomised rats replaced with constant infusions of CORT show a blunted ACTH and behavioural response to a noise stress, whereas animals replaced with the same dose of CORT but in a physiological pulsatile pattern have normal responses (Sarabdjitsingh et al. 2010). Interestingly, these changes are associated with brain-specific differences in c-fos activation, most particularly in the amygdala, suggesting that different brain circuits respond differentially to different patterns of CORT pulsatility. This study also demonstrated a phase-dependent and region-specific response, revealing a different response during the ascending and descending phases of each CORT pulse. This finding is complementary to data from Haller et al. (2000a, b), who found that rats exposed to male intruders during a rising phase of an endogenous CORT pulse were more aggressive than rats exposed to the same stimulus on a falling phase.

The relevance of CORT pulsatility in man needs investigation. We do know that the use of non-pulsatile oral hydrocortisone replacement therapy in patients with Addisons's disease is associated with a doubling in mortality (Bergthorsdottir et al. 2006) as well as increased morbidity predominantly related to mental and physical fatigue (Løvås et al. 2002). We have now designed a technique to provide physiological CORT replacement (Russell et al. 2014) and will be using it to investigate the importance of pulsatility for optimal cognitive and metabolic function.

Conclusion

Oscillatory activity is widespread in both our physical and biological environment. At the biological level, it can occur in multiple time domains. Within the HPA there is a very rapid (seconds) interaction at the level of GR:chromatin interactions, a slower (minutes) interaction between GR and its associated chaperones, an hourly cycle that emerges as a natural consequence of subhypothalamic feedforward: feedback interactions, and a daily oscillation regulated by the SCN. The circadian variation in CORT is actually made up of changes in the amplitude of the underlying subhypothalamic ultradian rhythm. The ultradian rhythm provides digital signals for both ACTH signalling to the adrenal and CORT signalling to tissues across the whole body, including the nervous, cardiovascular, metabolic and immune systems. The body has adapted to read these signals in a tissue-specific manner, allowing one hormone to have many effects in different tissues. This strategy of using a feedforward:feedback created digital signalling system is not unique to the HPA axis and is in fact commonly used across the endocrine system (Lightman and Terry 2014).

References

Arakane F, King SR, Du Y, Kallen CB, Walsh LP, Watari H, Stocco DM, Strauss JF 3rd (1997) Phosphorylation of steroidogenic acute regulatory protein (StAR) modulates its steroidogenic activity. J Biol Chem 272:32656–32662

Bergthorsdottir R, Leonsson-Zachrisson M, Odén A, Johannsson G (2006) Premature mortality in patients with Addison's disease: a population-based study. J Clin Endocrinol Metab 91 (12):4849–4853

Bhake RC, Leendertz JA, Linthorst AC, Lightman SL (2013) Automated 24-hours sampling of subcutaneous tissue free cortisol in humans. J Med Eng Technol 37:180–184

Biddie SC, Conway-Campbell BL, Lightman SL (2011) Dynamic regulation of glucocorticoid signalling in health and disease. Rheumatology 51:403–412

Boonen E, Langouche L, Janssens T, Meersseman P, Vervenne H, De Samblanx E, Pironet Z, Van Dijck L, Vander Perre S, Derese I, Van den Berghe G (2014) Impact of duration of critical illness on the adrenal glands of human intensive care patients. J Clin Endocrinol Metab 99 (11):4214–4222

Caraty A, Grino M, Locatelli A, Oliver C (1988) Secretion of corticotropin releasing factor (CRF) and vasopressin (AVP) into the hypophysial portal blood of conscious, unrestrained rams. Biochem Biophys Res Commun 155:841–849

Carnes M, Lent S, Feyzi J, Hazel D (1989) Plasma adrenocorticotropic hormone in the rat demonstrates three different rhythms within 24 h. Neuroendocrinology 50:17–25

Caron KM, Ikeda Y, Soo SC, Stocco DM, Parker KL, Clark BJ (1997) Characterization of the promoter region of the mouse gene encoding the steroidogenic acute regulatory protein. Mol Endocrinol 11:138–147

Conkright MD, Canettieri G, Screaton R, Guzman E, Miraglia L, Hogenesch JB, Montminy M (2003) TORCs. Mol Cell 12:413–423

Conway-Campbell BL, Sarabdjitsingh RA, McKenna MA, Pooley JR, Kershaw YM, Meijer OC, De Kloet ER, Lightman SL (2010) Glucocorticoid ultradian rhythmicity directs cyclical gene pulsing of the clock gene period 1 in rat hippocampus. J Neuroendocrinol 22:1093–1100

Conway-Campbell BL, George CL, Pooley JR, Knight DM, Norman MR, Hager GL, Lightman SL (2011) The HSP90 molecular chaperone cycle regulates cyclical transcriptional dynamics of the glucocorticoid receptor and its co-regulatory molecules CBP/P300 during ultradian ligand treatment. Mol Endocrinol 25(6):944–954

Conway-Campbell BL, Pooley JR, Hager GL, Lightman SL (2012) Molecular dynamics of ultradian glucocorticoid receptor action. Mol Cell Endocrinol 348:383–393 de Mendonca-Filho HT, Pereira KC, Fontes M, Vieira DA, de Mendonca ML, Campos LA, Castro-Faria-Neto HC (2006) Circulating inflammatory mediators and organ dysfunction after cardiovascular surgery with cardiopulmonary bypass: a prospective observational study. Crit Care 10:R46

Dickmeis T (2009) Glucocorticoids and the circadian clock. J Endocrinol 200(1):3–22

Droste SK, de Groote L, Atkinson HC, Lightman SL, Reul JM, Linthorst AC (2008) Corticosterone levels in the brain show a distinct ultradian rhythm but a delayed response to forced swim stress. Endocrinology 149:3244–3253

Einstein A (1905) Investigations on the theory of Brownian movement. Ann Phys 17:549

Engler D, Pham T, Liu JP, Fullerton MJ, Clarke IJ, Funder JW (1990) Studies of the regulation of the hypothalamic-pituitary-adrenal axis in sheep with hypothalamic-pituitary disconnection. II Evidence for in vivo ultradian hypersecretion of proopiomelanocortin peptides by the isolated anterior and intermediate pituitary. Endocrinology 127:1956–1966

Evanson NK, Tasker JG, Hill MN, Hillard CJ, Herman JP (2010) Fast feedback inhibition of the HPA axis by glucocorticoids is mediated by endocannabinoid signalling. Endocrinology 151 (10):4811–4819

Fulkerson WJ (1978) Synchronous episodic release of cortisol in the sheep. J Endocrinol 79:131–132

Gibbison B, Spiga F, Walker JJ, Russell GM, Stevenson K, Kershaw Y, Zhao Z, Henley D, Angelini GD, Lightman SL (2014) Dynamic pituitary-adrenal interactions in response to cardiac surgery. Crit Care Med 43:791–800

Hager GL, Elbi C, Johnson TA, Voss T, Nagaich AK, Schiltz RL, Qiu Y, John S (2006) Chromatin dynamics and the evolution of alternate promoter states. Chromosome Res 14(1):107–116

Haller J, Halasz J, Mikics E, Kruk MR, Makara GB (2000a) Ultradian corticosterone rhythm and the propensity to behave aggressively in male rats. J Neuroendocrinol 12:937–940

Haller J, Millar S, van de Schraaf J, de Kloet RE, Kruk MR (2000b) The active phase-related increase in corticosterone and aggression are linked. J Neuroendocrinol 12:431–436

Henley DE, Leendertz JA, Russell GM, Wood SA, Taheri S, Woltersdorf WW, Lightman SL (2009) Development of an automated blood sampling system for use in humans. J Med Eng Technol 33:199–208

Hill MN, Tasker JG (2012) Endocannabinoid signaling, glucocorticoid-mediated negative feedback, and regulation of the hypothalamic-pituitary-adrenal axis. Neuroscience 204:5–16

Hinz B, Hirschelmann R (2000) Rapid non-genomic feedback effects of glucocorticoids on CRF-induced ACTH secretion in rats. Pharm Res 17:1273–1277

Holaday JW, Martinez HM, Natelson BH (1977) Synchronized ultradian cortisol rhythms in monkeys: persistence during corticotropin infusion. Science 198:56–58

Ixart G, Barbanel G, Nouguier-Soule J, Assenmacher I (1991) A quantitative study of the pulsatile parameters of CRH-41 secretion in unanesthetized free-moving rats. Exp Brain Res 87:153–158

Jasper MS, Engeland WC (1991) Synchronous ultradian rhythms in adrenocortical secretion detected by microdialysis in awake rats. Am J Physiol 261:R1257–R1268

Jones MT, Brush FR, Neame RL (1972) Characteristics of fast feedback control of corticotrophin release by corticosteroids. J Endocrinol 55:489–497

Karst H, Berger S, Turiault M, Tronche F, Schütz G, Joëls M (2005) Mineralocorticoid receptors are indispensable for nongenomic modulation of hippocampal glutamate transmission by corticosterone. Proc Natl Acad Sci USA 102(52):19204–19207

Karst H, Berger S, Erdmann G, Schütz G, Joëls M (2010) Metaplasticity of amygdalar responses to the stress hormone corticosterone. Proc Natl Acad Sci USA 107(32):14449–14454

Lahat N, Zlotnick AY, Shtiller R, Bar I, Merin G (1992) Serum levels of IL-1, IL-6 and tumour necrosis factors in patients undergoing coronary artery bypass grafts or cholecystectomy. Clin Exp Immunol 89:255–260

Lightman SL, Conway-Campbell BL (2010) The crucial role of pulsatile activity of the HPA axis for continuous dynamic equilibration. Nat Rev Neurosci 11(10):710–718

Lightman SL, Terry JR (2014) The importance of dynamic signalling for endocrine regulation and drug development: relevance for glucocorticoid hormones. Lancet Diabetes Endocrinol 2 (7):593–599

Løvås K, Loge JH, Husebye ES (2002) Subjective health status in Norwegian patients with Addison's disease. Clin Endocrinol (Oxf) 56(5):581–588

McMaster A, Jangani M, Sommer P, Han N, Brass A, Beesley S, Lu W, Berry A, Loudon A, Donn R, Ray DW (2011) Ultradian cortisol pulsatility encodes a distinct, biologically important signal. PLoS One 6(1):e15766

Mershon JL, Sehlhorst CS, Rebar RW, Liu JH (1992) Evidence of a corticotropin-releasing hormone pulse generator in the macaque hypothalamus. Endocrinology 130:2991–2996

Mountjoy KG, Mortrud MT, Low MJ, Simerly RB, Cone RD (1994) Localization of the melanocortin-4 receptor (MC4-R) in neuroendocrine and autonomic control circuits in the brain. Mol Endocrinol 8:1298–1308

Qian X, Droste SK, Lightman SL, Reul JM, Linthorst AC (2012) Circadian and ultradian rhythms of free glucocorticoid hormone are highly synchronized between the blood, the subcutaneous tissue, and the brain. Endocrinology 153:4346–4353

Roth-Isigkeit A, Borstel TV, Seyfarth M, Schmucker P (1999) Perioperative serum levels of tumour-necrosis-factor alpha (TNF-alpha), IL-1 beta, IL-6, IL-10 and soluble IL-2 receptor in patients undergoing cardiac surgery with cardiopulmonary bypass without and with correction for haemodilution. Clin Exp Immunol 118:242–246

Rotsztejn W, Lalonde J, Normand M, Fortier C (1975) Feedback inhibition of adrenocorticotropin release by corticosterone infusions in the adrenalectomized rat. Can J Physiol Pharmacol 53:475–478

Russell GM, Henley DE, Leendertz J, Douthwaite JA, Wood SA, Stevens A, Woltersdorf WW, Peeters BW, Ruigt GS, White A, Veldhuis JD, Lightman SL (2010) Rapid glucocorticoid receptor-mediated inhibition of hypothalamic-pituitary-adrenal ultradian activity in healthy males. J Neurosci 30:6106–6115

Russell GM, Durant C, Ataya A, Papastathi C, Bhake R, Woltersdorf W, Lightman S (2014) Subcutaneous pulsatile glucocorticoid replacement therapy. Clin Endocrinol (Oxf) 81 (2):289–293

Sarabdjitsingh RA, Conway-Campbell BL, Leggett JD, Waite EJ, Meijer OC, de Kloet ER, Lightman SL (2010) Stress responsiveness varies over the ultradian glucocorticoid cycle in a brain-region-specific manner. Endocrinology 151:5369–5379

Sarabdjitsingh RA, Jezequel J, Pasricha N, Mikasova L, Kerkhofs A, Karst H, Groc L, Joëls M (2014) Ultradian corticosterone pulses balance glutamatergic transmission and synaptic plasticity. Proc Natl Acad Sci USA 111(39):14265–14270

Song KH, Park JI, Lee MO, Soh J, Lee K, Choi HS (2001) LH induces orphan nuclear receptor Nur77 gene expression in testicular Leydig cells. Endocrinology 142:5116–5123

Song KH, Park YY, Park KC, Hong CY, Park JH, Shong M, Lee K, Choi HS (2004) The atypical orphan nuclear receptor DAX-1 interacts with orphan nuclear receptor Nur77 and represses its transactivation. Mol Endocrinol 18:1929–1940

Spiga F, Lightman SL (2014) Dynamics of adrenal steroidogenesis in health and disease. Mol Cell Endocrinol 408:227–234

Spiga F, Waite EJ, Liu Y, Kershaw YM, Aguilera G, Lightman SL (2011) ACTH-dependent ultradian rhythm of corticosterone secretion. Endocrinology 152:1448–1457

Stavreva DA, Wiench M, John S, Conway-Campbell BL, McKenna MA, Pooley JR, Johnson TA, Voss TC, Lightman SL, Hager GL (2009) Ultradian hormone stimulation induces glucocorticoid receptor-mediated pulses of gene transcription. Nat Cell Biol 11:1093–1102

Sugawara T, Holt JA, Kiriakidou M, Strauss JF 3rd (1996) Steroidogenic factor 1-dependent promoter activity of the human steroidogenic acute regulatory protein (StAR) gene. Biochemistry 35:9052–9059

Takemori H, Kanematsu M, Kajimura J, Hatano O, Katoh Y, Lin XZ, Min L, Yamazaki T, Doi J, Okamoto M (2007) Dephosphorylation of TORC initiates expression of the StAR gene. Mol Cell Endocrinol 265–266:196–204

Vrang N, Larsen PJ, Møller M, Mikkelsen JD (1995) Topographical organization of the rat suprachiasmatic-paraventricular projection. J Comp Neurol 353(4):585–603

Waite EJ, McKenna M, Kershaw Y, Walker JJ, Cho K, Piggins HD, Lightman SL (2012) Ultradian corticosterone secretion is maintained in the absence of circadian cues. Eur J Neurosci 36:3142–3150

Walker JJ, Terry JR, Lightman SL (2010) Origin of ultradian pulsatility in the hypothalamic-pituitary-adrenal axis. Proc Biol Sci 277:1627–1633

Walker JJ, Spiga F, Waite E, Zhao Z, Kershaw Y, Terry JR, Lightman SL (2012) The origin of glucocorticoid hormone oscillations. PLoS Biol 10:e1001341

Walker JJ, Spiga F, Gupta R, Zhao Z, Lightman SL, Terry J (2014) Rapid intra-adrenal feedback regulation of glucocorticoid synthesis. J Roy Soc Interface. Published 12 Nov 2014. doi:10.1098/rsif.2014.0875 (J Roy Soc Interface 2015 12(102):20140875)

Weitzman ED, Fukushima D, Nogeire C, Roffwarg H, Gallagher TF, Hellman L (1971) Twenty-four hour pattern of the episodic secretion of cortisol in normal subjects. J Clin Endocrinol Metab 33:14–22

Whitehead G, Jo J, Hogg EL, Piers T, Kim DH, Seaton G, Seok H, Bru-Mercier G, Son GH, Regan P, Hildebrandt L, Waite E, Kim BC, Kerrigan TL, Kim K, Whitcomb DJ, Collingridge GL, Lightman SL, Cho K (2013) Acute stress causes rapid synaptic insertion of Ca2+-permeable AMPA receptors to facilitate long-term potentiation in the hippocampus. Brain 136(Pt 12):3753–3765

Windle RJ, Wood SA, Shanks N, Lightman SL, Ingram CD (1998) Ultradian rhythm of basal corticosterone release in the female rat: dynamic interaction with the response to acute stress. Endocrinology 139:443–450

The Genetics of Autism Spectrum Disorders

Guillaume Huguet, Marion Benabou, and Thomas Bourgeron

Abstract In the last 30 years, twin studies have indicated a strong genetic contribution to Autism Spectrum Disorders (ASD). The heritability of ASD is estimated to be 50 %, mostly captured by still unknown common variants. In approximately 10 % of patients with ASD, especially those with intellectual disability, *de novo* copy number or single nucleotide variants affecting clinically relevant genes for ASD can be identified. Given the function of these genes, it was hypothesized that abnormal synaptic plasticity and failure of neuronal/synaptic homeostasis could increase the risk of ASD. In parallel, abnormal levels of blood serotonin and melatonin were reported in a subset of patients with ASD. These biochemical imbalances could act as risk factors for the sleep/circadian disorders that are often observed in individuals with ASD. Here, we review the main pathways associated with ASD, with a focus on the roles of the synapse and the serotonin-NAS-melatonin pathway in the susceptibility of ASD.

G. Huguet • M. Benabou
Institut Pasteur, Human Genetics and Cognitive Functions Unit, Paris, France

CNRS UMR3571 Genes, Synapses and Cognition, Institut Pasteur, Paris, France

Sorbonne Paris Cité, Human Genetics and Cognitive Functions, University Paris Diderot, Paris, France

T. Bourgeron (✉)
Institut Pasteur, Human Genetics and Cognitive Functions Unit, Paris, France

CNRS UMR3571 Genes, Synapses and Cognition, Institut Pasteur, Paris, France

Sorbonne Paris Cité, Human Genetics and Cognitive Functions, University Paris Diderot, Paris, France

FondaMental Foundation, Créteil, France

Gillberg Neuropsychiatry Centre, Sahlgrenska Academy, University of Gothenburg, Gothenburg, Sweden
e-mail: thomasb@pasteur.fr

© The Author(s) 2016
P. Sassone-Corsi, Y. Christen (eds.), *A Time for Metabolism and Hormones*,
Research and Perspectives in Endocrine Interactions,
DOI 10.1007/978-3-319-27069-2_11

Introduction

Autism Spectrum Disorders (ASD) are a group of neuropsychiatric disorders characterized by problems in social communication as well as the presence of restricted interests and stereotyped and repetitive behaviors (Kanner 1943; Asperger 1944; Coleman and Gillberg 2012). Epidemiological studies estimate that more than 1 % of the population could receive a diagnosis of ASD (Elsabbagh et al. 2012; Developmental Disabilities Monitoring Network Surveillance Year Principal 2014). Individuals with ASD can also suffer from other psychiatric and medical conditions, including intellectual disability (ID), epilepsy, motor control difficulties, Attention-Deficit Hyperactivity Disorder (ADHD), tics, anxiety, sleep disorders, epilepsy, depression or gastrointestinal problems (Gillberg 2010; Moreno-De-Luca et al. 2013). The term ESSENCE, for 'Early Symptomatic Syndromes Eliciting Neurodevelopmental Clinical Examinations,' was coined by Christopher Gillberg to take into account this clinical heterogeneity and syndrome overlap (Gillberg 2010). There are four to eight times more males than females with ASD (Elsabbagh et al. 2012), but the sex ratio is more balanced in patients with ID and/or dysmorphic features (Miles et al. 2005). Autism can be studied as a category (affected vs. unaffected) or as a quantitative trait using auto- or hetero-questionnaires such as the Social Responsiveness Scale (SRS) or the autism quotient (AQ) (Ronald et al. 2006; Skuse et al. 2009; Constantino 2011). Using these tools, autistic traits seem to be normally distributed in clinical cases as well as in the general population (Ronald et al. 2006; Skuse et al. 2009; Constantino 2011).

The causes of autism remain largely unknown, but twin studies have constantly shown a high genetic contribution to ASD. Molecular genetics studies have identified more than 100 ASD risk genes carrying rare and penetrant deleterious mutations in approximately 10–25 % of patients (Huguet et al. 2013; Gaugler et al. 2014; Bourgeron 2015). In addition, quantitative genetics studies have shown that common genetic variants could capture almost all the heritability of ASD (Huguet et al. 2013; Gaugler et al. 2014). The genetic landscape of ASD is shaped by a complex interplay between common and rare variants and is most likely different from one individual to another (Gardener et al. 2011; Hallmayer et al. 2011; Bourgeron 2015). Remarkably, the susceptibility genes seem (Huguet et al. 2013) to converge in a limited number of biological pathways, including chromatin remodeling, protein translation, actin dynamics and synaptic functions (Bourgeron 2009; Toro et al. 2010; Huguet et al. 2013; Bourgeron 2015). In addition, several studies have pointed to a dysfunction of the serotonin-NAS-melatonin pathway in patients with ASD. Abnormalities of this pathway might increase the risk of circadian/sleep disorders often observed in patients with ASD.

In this chapter, we will detail the advances in the genetics of ASD (Fig. 1) with a focus on the role of both synapses and biological rhythms in the susceptibility of ASD (Abrahams and Geschwind 2008; Bourgeron 2009; Toro et al. 2010; Devlin and Scherer 2012; Huguet et al. 2013).

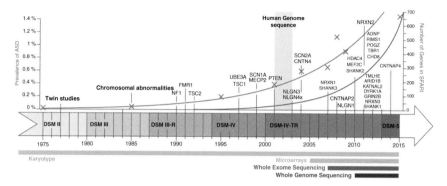

Fig. 1 The history of the genetics of autism from 1975 to 2015. The increase in the identified genes associated with ASD (SFARI—March 2015) is represented together with the prevalence of ASD (data taken from the Center for Disease Control and Prevention), the different versions of the Diagnostic Statistical Manual (from DSM II to DSM 5.0) and the advance in genetics technology (Adapted from Huguet and Bourgeron 2016)

Twin and Family Studies in ASD

Based on more than 13 twin studies published between 1977 and 2015, researchers have estimated the genetic and environmental contribution to ASD (Fig. 2). In 1977, the first twin study of autism by Folstein and Rutter (1977) reported on a cohort of 11 monozygotic (MZ) twins and 10 dizygotic (DZ) twins. This study showed that MZ twins were more concordant for autism—36 % (4/11)—compared with 0 % (0/10) for DZ twins. When a "broader autism phenotype" was used, the concordance increased to 92 % for MZ twins and to 10 % DZ twins (Bailey et al. 1995). Since this first small scale study, twin studies have constantly reported a higher concordance for ASD in MZ compared with DZ (Ritvo et al. 1985; Steffenburg et al. 1989; Bailey et al. 1995; Le Couteur et al. 1996). Between 2005 and 2009, three twin studies with relatively large groups of twins (285–3419) have reported high concordances for ASD in MZ twins (77–95 %) compared with DZ twins (31 %; Ronald et al. 2005; Taniai et al. 2008; Rosenberg et al. 2009). Notably, MZ concordances were similar to those reported in the previous studies, but DZ concordances were higher. In 2010, Lichtenstein et al. reported a relatively low concordance for ASD in 39 % of the MZ twins compared with other studies (the concordance for DZ twins in this study was 15 %). However, as previously indicated by studies using the "broader autism phenotype," all discordant MZ twins of this cohort had symptoms of ESSENCE (e.g., ID, ADHD, language delay, etc.). A significant proportion of the genetic contribution to ASD was shown to be shared with other neurodevelopmental disorders such as ADHD (>50 %) and learning disability (>40 %; Lichtenstein et al. 2010; Ronald et al. 2010; Lundstrom et al. 2011; Ronald and Hoekstra 2011). In summary, when all twin studies are taken into account, concordance for ASD is roughly 45 % for

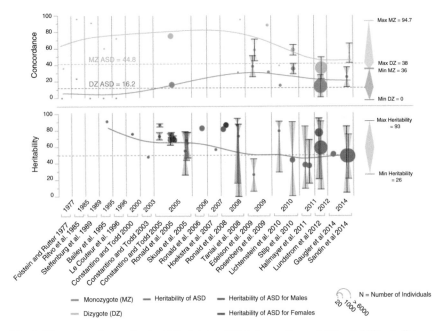

Fig. 2 The main twins studies in ASD. A total of 13 twins studies and 17 heritability studies are depicted. Means of concordance and heritability weighted by sample size are presented on the *right* of the figure (Adapted from Huguet and Bourgeron 2016)

MZ twins and 16 % for DZ twins (Ritvo et al. 1985; Steffenburg et al. 1989; Bailey et al. 1995; Le Couteur et al. 1996).

Family studies also showed that the recurrence of having a child with ASD increases with the proportion of the genome that the individual shares with one affected sibling or parent (Constantino et al. 2010; Risch et al. 2014; Sandin et al. 2014). In a population-based sample of 14,516 children diagnosed with ASD (Sandin et al. 2014), the relative risk for ASD (compared to the general population) was estimated to be 153.0 [95 % confidence interval (CI): 56.7–412.8] for MZ twins, 8.2 (3.7–18.1) for DZ twins; 10.3 (9.4–11.3) for full siblings, 3.3 (95 % CI, 2.6–4.2) for maternal half siblings, 2.9 (95 % CI, 2.2–3.7) for paternal half siblings, and 2.0 (95 % CI, 1.8–2.2) for cousins.

Heritability is the proportion of the phenotypic variation in a trait of interest, measured in a given studied population and in a given environment, that is co-varying with genetic differences among individuals in the same population. In 1995, based on a twin study, Bailey et al. estimated the heritability of autism to be 91–93 %. Since then, the estimation of heritability has differed from one study to another, but the genetic variance has accounted for at least 38 % and up to 90 % of the phenotypic variance (Hallmayer et al. 2011; Ronald and Hoekstra 2011; Sandin et al. 2014). Using a large cohort of 14,516 children diagnosed with ASD Sandin et al. (2014), estimated the heritability to be 0.50 (95 % CI, 0.45–0.56) and the non-shared environmental influence was also 0.50 (95 % CI, 0.44–0.55).

Surprisingly, only the additive genetic component and the non-shared environment seemed to account for the risk of developing ASD (Sandin et al. 2014).

In summary, epidemiological studies provide crucial information about the heritability of ASD. However, they do not inform us about the genes involved or the number and frequency of their variants. In the last 15 years, candidate genes and whole-genome analyses have been performed to address these questions.

From Chromosomal Rearrangements to Copy Number Variants in ASD

The first genetic studies that associated genetic variants with ASD used observations from cytogenetic studies (Gillberg and Wahlstrom 1985). However, because of the low resolution of the karyotypes (several Mb), it was almost impossible to associate a specific gene to ASD using this approach. The prevalence of large chromosomal abnormalities is estimated to be less than 2 % (Vorstman et al. 2006). Thanks to progress in molecular technologies such as Comparative Genomic Hybridization (CGH) or SNP arrays, the resolution in the detection of genomic imbalances has dramatically increased. Depending on the platforms, Copy Number Variants (CNVs) of more than 50 kb are now robustly detected (Pinto et al. 2011). Since the first articles published in 2006, a very large number of studies have investigated the contribution of CNVs to ASD (Jacquemont et al. 2006; Sebat et al. 2007). Several studies using the Simons Simplex Collection could even provide an estimation of the frequency of the *de novo* CNVs in patients with ASD compared with their unaffected siblings (Sanders et al. 2011). All together, *de novo* CNVs are present in 4–7 % of the patients with ASD compared to 1–2 % in the unaffected siblings and controls (Glessner et al. 2009; Sanders et al. 2011; Pinto et al. 2014). The studies have also indicated that *de novo* CNVs identified in patients are most likely altering genes and most especially genes associated with synaptic functions and/or regulated by FMRP, the protein responsible for the fragile X syndrome (Pinto et al. 2010, 2014). Beyond ASD, large CNVs (>400 kb) affecting exons are present in 15 % of patients with Developmental Delay (DD) or ID (Cooper et al. 2011). Most of the CNVs are private to each individual, but some are recurrently observed in independent patients. For example, three loci on chromosomal regions 7q11, 15q11.2–13.3, and 16p11.2 have been strongly associated with ASD (Ballif et al. 2007; Kumar et al. 2008; Weiss et al. 2008; Szafranski et al. 2010; Sanders et al. 2011; Leblond et al. 2014).

In summary, large chromosomal rearrangements and CNVs increase the risk of having ASD in 5–10 % of the individuals (Vorstman et al. 2006; Pinto et al. 2010, 2014). To go further in the identification of the ASD risk genes, candidate genes and whole exome/genomes studies were performed.

From Candidate Genes to Whole Exome/Genome Sequencing Studies in ASD

The first approach to associate a gene with ASD was to select specific candidate genes based on data coming from functional or genetic studies or a combination of the two. This approach was successful in identifying several synaptic genes associated with ASD such as *NLGN3, NLGN4X, SHANK3* and *NRXN1* (Jamain et al. 2003; Durand et al. 2007; Szatmari et al. 2007). Thanks to the advance in Next Generation Sequencing (NGS), we can now interrogate all genes of the genome in an unbiased manner using Whole Exome/Genome Sequencing (WES, WGS).

To date, more than 18 WES studies of sporadic cases of ASD (O'Roak et al. 2011, 2012a; Chahrour et al. 2012; Iossifov et al. 2012; Neale et al. 2012; Sanders et al. 2012; He et al. 2013; Lim et al. 2013; Liu et al. 2013, 2014; Willsey et al. 2013; Yu et al. 2013; An et al. 2014; De Rubeis et al. 2014; Iossifov et al. 2014; Samocha et al. 2014; Chang et al. 2015; Krumm et al. 2015) have been performed, comprising altogether more than >4000 families (Table 1). In almost all these studies, the authors have especially focused their analysis on the contribution of *de novo* Single Nucleotide Variants (SNVs) in ASD. All together, the average number of *de novo* coding SNVs per individual (including missense, splicing, frameshift, and stop-gain variants) is estimated to be approximately 0.86 in female patients, 0.73 in male patients, and 0.60 in unaffected male and female siblings (Krumm et al. 2014; Ronemus et al. 2014). Interestingly, *de novo* SNVs were three times more likely to be on the paternal chromosome than on the maternal one (Kong et al. 2012; O'Roak et al. 2012a) with an increase of almost two *de novo* mutations per year and doubled every 16.5 years (Kong et al. 2012).

Based on these studies (Iossifov et al. 2012; Neale et al. 2012; O'Roak et al. 2012a; Sanders et al. 2012), 3.6–8.8 % of the patients were shown to carry a *de novo* causative mutation (Iossifov et al. 2012) with a twofold increase of deleterious mutations in the patients compared with their unaffected siblings. In a meta-analysis, using more than 2500 families, Iossifov et al. (2014) found that de novo Likely Gene Disrupting (LGD) mutations (frameshift, nonsense and splice site) were more frequent in patients with ASD compared with unaffected siblings ($P = 5 \times 10^{-7}$). The carriers of these *de novo* LGDs were more likely diagnosed with a low non-verbal IQ. The *de novo* LGDs are significantly enriched in genes involved in chromatin modeling factors ($P = 4 \times 10^{-6}$) and in genes regulated by the FMRP complex ($p = 4 \times 10^{-7}$). Following these whole exome studies, targeted re-sequencing studies of the most compelling candidate genes were performed (O'Roak et al. 2012b). All together, 10 genes carrying *de novo* mutations were significantly associated with ASD: *CHD8, DYRK1A, GRIN2B, KATNAL2, RIMS1, SCN2A, POGZ, ADNP, ARID1B* and *TBR1*.

Only a few studies have analyzed the contribution of inherited SNVs in ASD. In 2013, Lim et al. analyzed whole exome sequencing of 933 cases (ASD) and 869 controls for the presence of rare complete human knockouts (KO) with homozygous or compound heterozygous loss-of-function (LoF) variants (≤ 5 %

Table 1 Summary of the main whole exome/genome sequencing studies in ASD

Studies	Tech	#ASD analyzed in the study	# ASD specific to this study	#ASD coming from other studies	#Controls	#Unaffected sibling	#Parents	Analysis of de novo variants	Analysis of inherited variants
O'Roak et al. (2011)	WES	20	20	–	–	20	38	X	–
O'Roak et al. (2012a)	WES	229	209	20 (O'Roak et al. 2011)	–	50	418	X	–
Neale et al. (2012)	WES	175	175	–	–	–	350	X	–
Sanders et al. (2012)	WES	238	238	–	–	200	476	X	–
Iossifov et al. (2012)	WES	343	343	–	–	343	686	X	X
Chahrour et al. (2012)	WES	16	16	–	–	–	–	–	X
Yu et al. (2013)	WES	401	163	238 (Iossifov et al. 2012; Neale et al. 2012; Sanders et al. 2012)	–	114	326	X	X
Lim et al. (2013)	WES	1496	1004	492 (Sanders et al. 2012)	5474	–	–	–	X
Liu et al. (2013)	WES	1039	–	1039 (O'Roak et al. 2011, 2012a; Iossifov et al. 2012; Neale et al. 2012; Sanders et al. 2012; Liu et al. 2013)	869	–	–	–	X
He et al. (2013)	WES	1867	–	1867 (O'Roak et al. 2011, 2012a; Iossifov et al. 2012; Neale et al. 2012; Sanders et al. 2012; Liu et al. 2013)	870	593	1870	X	X
Willsey et al. (2013)	WES	1099	56	1043 (O'Roak et al. 2011, 2012a; Iossifov et al. 2012; Neale et al. 2012; Sanders et al. 2012; Liu et al. 2013)	–	56	112	X	X

(continued)

Table 1 (continued)

Studies	Tech	#ASD analyzed in the study	# ASD specific to this study	#ASD coming from other studies	#Controls	#Unaffected sibling	#Parents	Analysis of *de novo* variants	Analysis of inherited variants
Liu et al. (2014)	WES	1967	–	1967 (O'Roak et al. 2011, 2012a; Iossifov et al. 2012; Kong et al. 2012; Neale et al. 2012; Sanders et al. 2012; Liu et al. 2013; Willsey et al. 2013)	870	593	2070	X	X
Samocha et al. (2014)	WES	1078	–	1078 (Asperger 1944; Coleman and Gillberg 2012; Developmental Disabilities Monitoring Network Surveillance Year Principal 2014; Elsabbagh et al. 2012; Gillberg 2010; Constantino 2011)	–	343	2156	X	–
Iossifov et al. (2014)	WES	2	1576	932 (O'Roak et al. 2011, 2012a; Iossifov et al. 2012; Neale et al. 2012; Sanders et al. 2012; Liu et al. 2013)	–	1911	5016	X	–
An et al. (2014)	WES	40	40	–	–	8	80	X	X
De Rubeis et al. (2014a)	WES	2270	–	2270 (O'Roak et al. 2011, 2012a; Iossifov et al. 2012, 2014; Neale et al. 2012; Sanders et al. 2012)	5397	–	4540	X	X
Chang et al. (2015)	WES	932	–	932 (Levy et al. 2011; O'Roak et al. 2011, 2012a; Iossifov et al. 2012; Sanders et al. 2012)	–	593	1580	X	X
Krumm et al. (2015)	WES	2377	–	2377 (O'Roak et al. 2011, 2012a; Iossifov et al. 2012, 2014; Neale et al. 2012; Sanders et al. 2012)	–	1786	4754	X	X

Kong et al. (2012)	WGS	40	44	–	–	7	–	136	X	–
Michaelson et al. (2012)	WGS	20	20	–	–	–	–	–	X	–
Shi et al. (2013)	WGS	1	1	–	–	6	–	2	X	X
Jiang et al. (2013)	WGS	32	32	–	–	–	–	64	X	X
Yuen et al. (2015)	WGS	85	85	32 (Jiang et al. 2013)	–	–	–	170	X	X
Nemirovsky et al. (2015)	WGS	1		–	–	–	–	–	–	X

frequency). They observed a significant twofold increase in complete KOs in patients with ASD compared to controls. They estimated that such complete KO mutations could account for 3 % of the patients with ASD. For the X chromosome, there was a significant 1.5-fold increase in complete KO in affected males compared to unaffected males that could account for 2 % of males with ASD (Lim et al. 2013). The same year, Yu et al. (2013) analyzed 104 consanguineous families including 79 families with a single child with ASD (simplex families) and 25 families with more than one affected individual (multiplex families) collected by the Homozygosity Mapping Collaborative for Autism (HMCA). They identified biallelic mutations in *AMT*, *PEX7*, *SYNE1*, *VPS13B*, *PAH*, and *POMGNT1*. Finally, a very recent study by Krumm et al. (2015) ascertained the relative impact of inherited and *de novo* variants (CNVs or SNVs) on ASD risk in 2377 families. Inherited truncating variants were enriched in probands (for SNV odds ratio = 1.14, P = 0.0002; for CNV odds ratio = 1.23, P = 0.001) in comparison to unaffected siblings (Krumm et al. 2015). Interestingly, they also observed a significant maternal transmission bias of inherited LGD to sons. New ASD-risk genes were also identified such as *RIMS1*, *CUL7* and *LZTR1*.

To date, few whole genome sequencing studies have been published for ASD (Table 1). Michaelson et al. (2012) analyzed 40 WGS of monozygotic twins concordant for ASD and their parents. They proposed that ASD-risk genes could be hot spots of mutation in the genome and confirmed the association between ASD and *de novo* mutations in *GPR98*, *KIRREL3* and *TCF4*. Shi et al. (2013) analyzed a large pedigree with two sons affected with ASD and six unaffected siblings, focusing on inherited mutations. They identified *ANK3* as the most likely candidate gene. In 2015, Yuen et al. analyzed 85 families with two children affected with ASD. This study represents the largest published WGS data set in ASD. They identified 46 ASD-relevant mutations present in 36 of 85 (42.4 %) families. Only 16 ASD-relevant mutations of 46 (35 %) identified were *de novo*. Very interestingly, for more than half of the families (69.4 %; 25 of 36), the two affected siblings did not share the same rare penetrant ASD risk variant(s).

Whole genome sequencing is also very efficient to identify mutations in regions of the human genome that are wrongly annotated and in regions that are highly GC rich. For example, mutations on the *SHANK3* gene were rarely identified using whole exome sequencing, given its high GC content (Leblond et al. 2014). In contrast, whole genome sequencing could successfully identify *SHANK3* mutations (Nemirovsky et al. 2015; Yuen et al. 2015).

The Common Variants in ASD

In the general population, one individual carries on average three million genetic variants in comparison to the reference human genome sequence (Xue et al. 2012; Fu et al. 2013; Genome of the Netherlands and Genome of the Netherlands 2014). The vast majority of the variants (>95 %) are the so-called common variants shared with more than 5 % of the human population (Xue et al. 2012; Fu et al. 2013; Genome of the Netherlands and Genome of the Netherlands 2014). While there is not a clear border between common and rare variants, it is nevertheless interesting to estimate the role of the genetic variants found in the general population in the susceptibility to ASD.

Using quantitative genetics, Klei et al. (2012) estimated that common variants were contributing to a high proportion of the liability of ASD: 40 % in simplex families and 60 % in multiplex families. In 2014, a study by Gaugler et al. used the same methodology (Yang et al. 2011) and provided an estimation of the heritability (52.4 %) that is almost exclusively due to common variation, leaving only 2.6 % of the liability to the rare variants. The contribution of common variants is therefore important, but unfortunately the causative SNPs still remain unknown since they are numerous (>1000) and each is associated with a low risk. To date, the largest genome wide association studies (GWAS) performed on <5000 families with ASD were underpowered to identify a single SNP with genome wide significance (Anney et al. 2012; Cross-Disorder Group of the Psychiatric Genomics 2013).

The recruitment of larger cohorts of patients with dimensional phenotypes is therefore warranted to better ascertain the heritability of ASD and to identify the genetic variants, which explain most of the genetic variance.

The Genetic Architecture of ASD

Based on the results obtained from epidemiological and molecular studies, it is now accepted that the genetic susceptibility to ASD can be different from one individual to another with a combination of rare deleterious variants (R) and a myriad of low-risk alleles [also defined as the genetic background (B)]. Most of the inherited part of ASD seems to be due to common variants observed in the general population, with only a small contribution from rare variants (Fig. 3). Importantly, while the *de novo* mutations are considered per se as genetic factors, they do not contribute to the heritability since they are only present in the patient (with the relatively rare exception of germinal mosaicisms present in one of the parental germlines and transmitted to multiple children). These *de novo* events could therefore be considered as "environmental causes" of ASD but acting on the DNA molecule. It is currently estimated that more than 500–1000 genes could account for these "monogenic forms" of ASD (Iossifov et al. 2012; Sanders et al. 2012), confirming the high degree of genetic heterogeneity.

The interplay between the rare or *de novo* variants R and the background B will also influence the phenotypic diversity observed in the patients carrying rare deleterious mutations. In some individuals, a genetic background B will be able to buffer or compensate the impact of the rare genetic variations R. In contrast, in some individuals, the buffering capacity of B will not be sufficient to compensate the impact of R and they will develop ASD (Rutherford 2000; Hartman et al. 2001). In the R'n'B model, ASD can be regarded as a collection of many genetic forms of

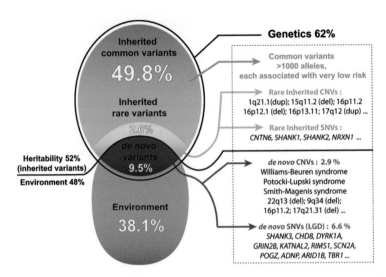

Fig. 3 Relative contribution of the genetics and environment in ASD. Based on twin and familial studies, it is estimated that the genetic and environmental contributions to ASD are approximately 50/50 %. Most of the heritable part seems to be due to common variants observed in the general population, with a small contribution of rare variants. Importantly, the *de novo* mutations are genetic causes of ASD but do not contribute to the heritability since there are only present in the patient. These *de novo* events are therefore considered to be "environmental causes" of ASD, but acting on the DNA molecule (Adapted from Huguet and Bourgeron 2016)

"autisms," each with a different etiology ranging from monogenic to polygenic models.

The presence of multiple hits of rare CNVs, SNVs or indels in a single individual also illustrates the complexity of the genetic landscape of ASD (Girirajan et al. 2010, 2012; Leblond et al. 2012). In addition, the analysis of the whole genome sequence of multiplex families also indicates that clinically relevant mutations can be different from one affected sib to another even in a single family (Yuen et al. 2015). It is therefore still difficult to ascertain robust genotype-phenotype relationships based on our current knowledge.

Fortunately, although the ASD-risk genes are numerous, they seem to converge in a limited number of biological pathways that are currently scrutinized by many researchers.

The Biological Pathways Associated with ASD

Unbiased pathway analyses indicated that ASD-risk genes seem to be enriched in groups of proteins with specific functions (Voineagu et al. 2011; De Rubeis et al. 2014; Ronemus et al. 2014; Uddin et al. 2014; Hormozdiari et al. 2015). Pinto et al. (2014) analyzed the burden of CNVs in 2446 individuals with ASD and

2640 controls and found enrichment in genes coding post-synaptic density proteins and FMRP targets. Ronemus et al. (2014) reviewed the results of four whole exome sequencing studies and showed an enrichment of mutated genes in chromatin modifier genes ($P = 4 \times 10^{-6}$) and FMRP targets ($P = 7 \times 10^{-6}$). Protein-protein interactions (PPI) analyses of the genes carrying LGD mutations also showed enrichment in proteins involved in neuronal development and axon guidance, signaling pathways and chromatin and transcription regulation. De Rubeis et al. (2014) also used the PPI network and showed enrichment in clusters of proteins involved in the cell junction TGF beta pathway, cell communication and synaptic transmission, neurodegeneration and transcriptional regulation.

In parallel to the genetic studies, several transcriptomic analyses were performed using post-mortem brain of individuals with ASD (Voineagu et al. 2011; Gupta et al. 2014). Several genes were differentially expressed or correlated between brain regions. Two network modules were identified. The first module was related to interneurons and to genes involved in synaptic function (downregulated in brains from ASD patients compared to controls). The second module was related to immunity and microglia activation (upregulated in brains from ASD patients compared to controls).

Based on these results, neurobiological studies using cellular and animal models have been performed to identify the main mechanisms leading to ASD. Remarkably, several studies showed that neuronal activity seems to regulate the function of many of the ASD-risk genes, leading to the hypothesis that abnormal synaptic plasticity and failure of neuronal/synaptic homeostasis could play a key role in the susceptibility to ASD (Belmonte and Bourgeron 2006; Auerbach et al. 2011; Toro et al. 2010). Here, we will only depict four main biological pathways associated with ASD: chromatin remodeling, protein synthesis, protein degradation, and synaptic function (Fig. 4). In parallel, several biochemical studies have indicated a dysfunction in the serotonin-NAS-melatonin pathway.

Chromatin Remodeling Mutations in genes encoding key regulators of chromatin remodeling and gene transcription (e.g., *MECP2, MEF2C, HDAC4, CHD8* and *CTNNB1*) have been reported in individuals with ASD (Fig. 4). Remarkably, a subset of these genes is regulated by neuronal activity and influences neuronal connectivity and synaptic plasticity (Cohen et al. 2011; Sando et al. 2012; Ebert et al. 2013).

Protein Synthesis The level of synaptic proteins can be influenced by neuronal activity through global and local synaptic mRNA translation (Ma and Blenis 2009). Several genes involved in such activity-driven regulation of synaptic proteins have been found to be mutated in individuals with ASD (Kelleher and Bear 2008). For example, the mTOR pathway controls global mRNA translation and its deregulation causes diseases associated with increased cell proliferation and loss of autophagy, including cancer (Ma and Blenis 2009), but also increases the risk for ASD. Remarkably, mutations in the repressor of the mTOR pathway such as *NF1, PTEN* and *SynGAP1* cause an increase of translation in neurons and at the synapse (Auerbach et al. 2011). Mutations of the FMRP–EIF4E–CYFIP1 complex cause the

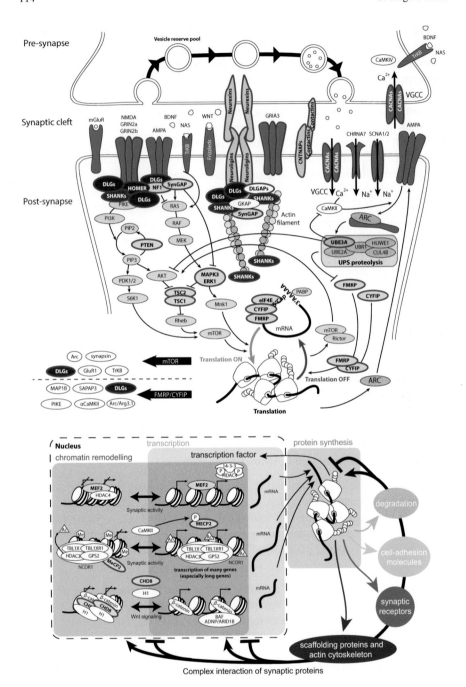

Fig. 4 Examples of the biological pathways associated with ASD. The ASD-risk genes code for proteins involved in chromatin remodeling, transcription, protein synthesis and degradation, cytoskeleton dynamics, and synaptic functions. Proteins associated with ASD are *circled* in *red* (Adapted from Huguet and Bourgeron 2016)

fragile X syndrome and increase the risk of ASD (Budimirovic and Kaufmann 2011). This protein complex controls local translation of mRNA at the synapse and acts downstream of the Ras-ERK signaling pathway. This complex regulates the translation of more than 1000 specific genes, many of which are ASD risk genes (De Rubeis et al. 2013; Fernandez et al. 2013; Gkogkas et al. 2013; Santini et al. 2013). An alteration of this FMRP–EIF4E–CYFIP1 complex should therefore create an imbalance in the level of many synaptic proteins that are associated with ASD.

Protein Degradation The Ubiquitin-Proteasome System (UPS) is central for the degradation of the proteins and, consequently, for the regulation of synapse composition, assembly and elimination (Mabb and Ehlers 2010). The *UBE3A* gene encodes a ubiquitin ligase, is mutated in patients with Angelman syndrome and is duplicated on the maternal chromosome 15q11 in individuals with ASD. Neuronal activity increases *UBE3A* transcription through the MEF2 complex and regulates excitatory synapse development by controlling the degradation of ARC, a synaptic protein that decreases long-term potentiation by promoting the internalization of AMPA receptors (Greer et al. 2010).

Synaptic Functions Many proteins encoded by ASD-risk genes participate in different aspects of neuronal connectivity, such as glutamatergic (e.g., GRIN2B), GABAergic (e.g., GABRA3 and GABRB3) and glycinergic (e.g., GLRA2) neurotransmission, neuritogenesis (e.g., CNTN), the establishment of synaptic identity (e.g., cadherins and protocadherins), neuronal conduction (CNTNAP2) and permeability to ions (CACNA1, CACNA2D3 and SCN1A). Some of these proteins are directly involved in activity-driven synapse formation, such as the neurexins (NRXNs) and neuroligins (NLGNs). Some are scaffolding proteins involved in the positioning of cell-adhesion molecules and neurotransmitter receptors at the synapse (Sheng and Kim 2011; Choquet and Triller 2013). For example, deletions, duplications and coding mutations in the three *SHANK* genes (*SHANK1, SHANK2* and *SHANK3*) have been recurrently reported in individuals with ASD (Leblond et al. 2014). SHANK proteins assemble into large molecular platforms in interaction with glutamate receptors and actin-associated proteins (Grabrucker et al. 2011). In vitro, *SHANK3* mutations identified in individuals with ASD reduce actin accumulation in spines affecting the development and morphology of dendrites as well as the axonal growth cone motility (Durand et al. 2012). In mice, mutations in *SHANK3* decrease spine density in the hippocampus but also increase dendritic arborizations in striatal neurons (Peca et al. 2011). Mice mutated in *SHANK* present with behavior resembling autistic features in humans. *Shank1* knockout mice display increased anxiety, decreased vocal communication, decreased locomotion and, remarkably, enhanced working memory, but decreased long-term memory (Hung et al. 2008; Silverman et al. 2011; Wohr et al. 2011). *Shank2* knockout mice present hyperactivity, increased anxiety, repetitive grooming, and abnormalities in vocal and social behaviors (Schmeisser et al. 2012; Won et al. 2012). *Shank3* knockout mice show self-injurious repetitive grooming, and deficits in social interaction and communication (Bozdagi et al. 2010; Peca et al. 2011; Wang et al. 2011; Yang et al. 2012).

The Serotonin-NAS-Melatonin Pathway

In parallel to the genetic investigations pointing to the biological pathways described above, several biological abnormalities have been reported in individuals with ASD, including neurochemical, immunological, endocrine or metabolic traits (Lam et al. 2006; Rossignol and Frye 2012), which may provide insights into the pathophysiology of ASD. Among these, elevated blood serotonin is one of the most replicated findings (Pagan et al. 2012) (Fig. 5). A deficit in melatonin (which chemically derives from serotonin) in the blood or urine of individuals with ASD has also been described in several studies (Tordjman et al. 2005; Melke et al. 2008) and is associated with increased peripheral *N*-acetylserotonin (NAS), the intermediate metabolite between serotonin and melatonin. Several mutations affecting this pathway were identified but the mechanisms of these biochemical impairments remain mostly unexplained. Melatonin is a neurohormone mainly synthesized in

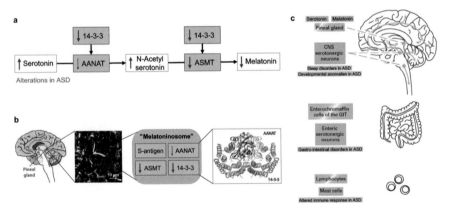

Fig. 5 The serotonin-NAS-melatonin pathway in ASD. (**a**) The serotonin-NAS-melatonin synthesis pathway consists of two enzymatic steps involving first the *N*-acetylation of serotonin to *N*-acetylserotonin (NAS) by the rate-limiting enzyme AANAT and the methylation of NAS by the ASMT (also known as HIOMT). Different alterations such as higher serotonin or low melatonin levels were observed in the blood of patients with ASD. The enzymes are represented in *gray* and metabolites are in *white*. Alterations of the biochemical parameters are shown with *red arrows*. (**b**) A schematic view of the pineal gland with the pinealocytes that contain the "melatoninosome." This complex includes at least four proteins: AANAT, ASMT, 14-3-3 and S-antigen. The immunofluorescence confocal image of AANAT (*green*) and 14-3-3δ/ε (*red*) in the pinealocytes is adapted from Maronde et al. (2011). The structure of the 14-3-3ζ homodimer binding to AANAT is adapted from Obsil et al. (2001). (**c**) Main sources of serotonin (*blue*) and melatonin (*green*) and the symptoms or comorbidities of ASD relevant to alterations in serotonin and melatonin levels observed in ASD (Adapted from Pagan et al. 2012)

the pineal gland during the night. It is a biological signal of light/dark cycles and is considered to be a major circadian synchronizer. It is also a modulator of metabolism, immunity, reproduction and digestive function. Furthermore, it displays antioxidant and neuroprotective properties and can directly modulate neuronal networks (Bourgeron 2007). Melatonin appears as a therapeutic target of the frequently reported sleep disorders associated with ASD (Andersen et al. 2008; Wright et al. 2011; Malow et al. 2012). NAS displays intrinsic biological properties: it is an agonist of the TrkB receptor and may thus share the neurotrophic properties of brain-derived neurotrophic factor (BDNF), the canonical TrkB ligand (Jang et al. 2010; Sompol et al. 2011). Serotonin conversion into melatonin involves two sequential enzymatic steps: N-acetylation of serotonin into NAS by arylalkylamine N-acetyltranferase (AANAT, EC: 2.3.1.87) followed by methylation by acetylserotonin O-methyltransferase (ASMT, also called hydroxyindole O-methyltransferase HIOMT, EC: 2.1.1.4) (Fig. 5a). We previously showed that deleterious mutations of the *ASMT* gene could disrupt melatonin synthesis in a subset of patients with ASD. Nevertheless, the frequency of such a deleterious mutation is too low (2 % of the cases) to explain the relatively high frequency of melatonin deficit in ASD (>50 % of the patients, taking as a threshold the fifth percentile of the controls). More recently, we observed a low level of the 14-3-3 proteins both in the blood platelet and pineal gland of patients with ASD (Pagan et al. 2014). These ubiquitous chaperone proteins are known to form a protein complex, the 'melatoninosome,' involving AANAT and ASMT in pinealocytes (Obsil et al. 2001; Maronde et al. 2011). This interaction between 14-3-3 and AANAT and/or ASMT might be essential for the production of melatonin and an adequate balance of the serotonin-NAS-melatonin pathway. Indeed, a low level of 14-3-3 could eventually lead to a deficit in enzyme activity, contributing to the global disruption of the serotonin-NAS-melatonin pathway observed in ASD. Studies investigating the regulation of the complex 14-3-3/ASMT/AANAT in ASD and controls are in progress.

Perspectives

In the last 30 years, very significant progress has been made in the genetics of ASD. We now have a better knowledge on the genetic architecture of this heterogeneous syndrome and some of the biological pathways have been investigated using different approaches such as cellular and animal models. There are, however, many aspects of the genetics of ASD that remain largely unknown.

The first challenge concerns the role of the common variants. These variants are most likely playing a key role in the susceptibility to ASD and in the severity of the symptoms. But, because the impact of each single SNP is very low, it is currently impossible to identify the risk alleles using conventional GWAS. In human quantitative traits such as height, neuroanatomical diversity or intellectual quotient, very large cohorts of many thousands of individuals are necessary to identify the main causative SNPs (Toro et al. 2014; Yang et al. 2010; Deary et al. 2012).

The second challenge concerns the stratification of the patients and the role of the ASD-risk genes during brain development/function. Based on our current knowledge, the genetic architecture of ASD seems to be different from one individual to another, with possibly contrasting impact on when and where neuronal connectivity could be atypical compared to the general population. For example, in animal models, several mutations lead to higher connectivity whereas other mutations alter synaptic density. It is therefore crucial to increase our knowledge from a basic research perspective about the biological roles of the ASD-risk genes and their partners.

Finally, while we all agree that biological research is necessary to improve the quality of life of the patients and their families (for example, to alleviate the comorbidities associated with ASD like sleep and gastrointestinal problems), progress should also be made toward better recognition and inclusion of people with neuropsychiatric conditions in our societies (no mind left behind). Hopefully, increasing knowledge in genetics, neurology and psychology should allow for better diagnosis, care for and integration of individuals with autism.

Acknowledgments We thank Roberto Toro for critical reading of the manuscript. This work was funded by the Institut Pasteur, the Bettencourt-Schueller foundation, Centre National de la Recherche Scientifique, University Paris Diderot, Agence Nationale de la Recherche (SynDiv-ASD), the Conny-Maeva Charitable Foundation, the Cognacq Jay Foundation, the Orange Foundation, and the Fondamental Foundation.

References

Abrahams BS, Geschwind DH (2008) Advances in autism genetics: on the threshold of a new neurobiology. Nat Rev Genet 9:341–355

An JY, Cristino AS, Zhao Q, Edson J, Williams SM, Ravine D, Wray J, Marshall VM, Hunt A, Whitehouse AJ, Claudianos C (2014) Towards a molecular characterization of autism spectrum disorders: an exome sequencing and systems approach. Transl Psychiatry 4:e394

Andersen IM, Kaczmarska J, McGrew SG, Malow BA (2008) Melatonin for insomnia in children with autism spectrum disorders. J Child Neurol 23:482–485

Anney R et al (2012) Individual common variants exert weak effects on the risk for autism spectrum disorderspi. Hum Mol Genet 21:4781–4792

Asperger H (1944) Die "autistischen Psychopathen" im Kindesalter. Arch Psychiatr Nervenkr 177:76–137

Auerbach BD, Osterweil EK, Bear MF (2011) Mutations causing syndromic autism define an axis of synaptic pathophysiology. Nature 480:63–68

Bailey A, Le Couteur A, Gottesman I, Bolton P, Simonoff E, Yuzda E, Rutter M (1995) Autism as a strongly genetic disorder: evidence from a British twin study. Psychol Med 25:63–77

Ballif BC, Hornor SA, Jenkins E, Madan-Khetarpal S, Surti U, Jackson KE, Asamoah A, Brock PL, Gowans GC, Conway RL, Graham JM Jr, Medne L, Zackai EH, Shaikh TH, Geoghegan J, Selzer RR, Eis PS, Bejjani BA, Shaffer LG (2007) Discovery of a previously unrecognized microdeletion syndrome of 16p11.2-p12.2. Nat Genet 39:1071–1073

Belmonte MK, Bourgeron T (2006) Fragile X syndrome and autism at the intersection of genetic and neural networks. Nat Neurosci 9:1221–1225

Bourgeron T (2007) The possible interplay of synaptic and clock genes in autism spectrum disorders. Cold Spring Harb Symp Quant Biol 72:645–654

Bourgeron T (2009) A synaptic trek to autism. Curr Opin Neurobiol 19:231–234

Bourgeron T (2015) From the genetic architecture to synaptic plasticity in autism spectrum disorder. Nat Rev Neurosci 16:551–563

Bozdagi O, Sakurai T, Papapetrou D, Wang X, Dickstein DL, Takahashi N, Kajiwara Y, Yang M, Katz AM, Scattoni ML, Harris MJ, Saxena R, Silverman JL, Crawley JN, Zhou Q, Hof PR, Buxbaum JD (2010) Haploinsufficiency of the autism-associated Shank3 gene leads to deficits in synaptic function, social interaction, and social communication. Mol Autism 1:15

Budimirovic DB, Kaufmann WE (2011) What can we learn about autism from studying fragile X syndrome? Dev Neurosci 33:379–394

Chahrour MH, Yu TW, Lim ET, Ataman B, Coulter ME, Hill RS, Stevens CR, Schubert CR, Collaboration AAS, Greenberg ME, Gabriel SB, Walsh CA (2012) Whole-exome sequencing and homozygosity analysis implicate depolarization-regulated neuronal genes in autism. PLoS Genet 8:e1002635

Chang J, Gilman SR, Chiang AH, Sanders SJ, Vitkup D (2015) Genotype to phenotype relationships in autism spectrum disorders. Nat Neurosci 18:191–198

Choquet D, Triller A (2013) The dynamic synapse. Neuron 80:691–703

Cohen S, Gabel HW, Hemberg M, Hutchinson AN, Sadacca LA, Ebert DH, Harmin DA, Greenberg RS, Verdine VK, Zhou Z, Wetsel WC, West AE, Greenberg ME (2011) Genome-wide activity-dependent MeCP2 phosphorylation regulates nervous system development and function. Neuron 72:72–85

Coleman M, Gillberg C (2012) The autisms. Oxford University Press, New York

Constantino JN (2011) The quantitative nature of autistic social impairment. Pediatr Res 69:55R–62R

Constantino JN, Zhang Y, Frazier T, Abbacchi AM, Law P (2010) Sibling recurrence and the genetic epidemiology of autism. Am J Psychiatry 167:1349–1356

Cooper GM, Coe BP, Girirajan S, Rosenfeld JA, Vu TH, Baker C, Williams C, Stalker H, Hamid R, Hannig V, Abdel-Hamid H, Bader P, McCracken E, Niyazov D, Leppig K, Thiese H, Hummel M, Alexander N, Gorski J, Kussmann J, Shashi V, Johnson K, Rehder C, Ballif BC, Shaffer LG, Eichler EE (2011) A copy number variation morbidity map of developmental delay. Nat Genet 43:838–846

Cross-Disorder Group of the Psychiatric Genomics C (2013) Identification of risk loci with shared effects on five major psychiatric disorders: a genome-wide analysis. Lancet 381:1371–1379

De Rubeis S, Pasciuto E, Li KW, Fernandez E, Di Marino D, Buzzi A, Ostroff LE, Klann E, Zwartkruis FJ, Komiyama NH, Grant SG, Poujol C, Choquet D, Achsel T, Posthuma D, Smit AB, Bagni C (2013) CYFIP1 coordinates mRNA translation and cytoskeleton remodeling to ensure proper dendritic spine formation. Neuron 79:1169–1182

De Rubeis S, He X, Goldberg AP, Poultney CS, Samocha K, Cicek AE, Kou Y, Liu L, Fromer M, Walker S, Singh T, Klei L, Kosmicki J, Shih-Chen F, Aleksic B, Biscaldi M, Bolton PF, Brownfeld JM, Cai J, Campbell NG, Carracedo A, Chahrour MH, Chiocchetti AG, Coon H,

Crawford EL, Curran SR, Dawson G, Duketis E, Fernandez BA, Gallagher L, Geller E, Guter SJ, Hill RS, Ionita-Laza J, Jimenz Gonzalez P, Kilpinen H, Klauck SM, Kolevzon A, Lee I, Lei I, Lei J, Lehtimäki T, Lin CF, Ma'ayan A, Marshall CR, McInnes AL, Neale B, Owen MJ, Ozaki N, Parellada M, Parr JR, Purcell S, Puura K, Rajagopalan D, Rehnström K, Reichenberg A, Sabo A, Sachse M, Sanders SJ, Schafer C, Schulte-Rüther M, Skuse D, Stevens C, Szatmari P, Tammimies K, Valladares O, Voran A, Li-San W, Weiss LA, Willsey AJ, Yu TW, Yuen RK, DDD Study, Homozygosity Mapping Collaborative for Autism, UK10K Consortium, Cook EH, Freitag CM, Gill M, Hultman CM, Lehner T, Palotie A, Schellenberg GD, Sklar P, State MW, Sutcliffe JS, Walsh CA, Scherer SW, Zwick ME, Barett JC, Cutler DJ, Roeder K, Devlin B, Daly MJ, Buxbaum JD (2014) Synaptic, transcriptional and chromatin genes disrupted in autism. Nature 515:209–215
Deary IJ, Yang J, Davies G, Harris SE, Tenesa A, Liewald D, Luciano M, Lopez LM, Gow AJ, Corley J, Redmond P, Fox HC, Rowe SJ, Haggarty P, McNeill G, Goddard ME, Porteous DJ, Whalley LJ, Starr JM, Visscher PM (2012) Genetic contributions to stability and change in intelligence from childhood to old age. Nature 482:212–215
Developmental Disabilities Monitoring Network Surveillance Year Principal I (2014) Prevalence of autism spectrum disorder among children aged 8 years – autism and developmental disabilities monitoring network, 11 sites, United States, 2010. MMWR Surveill Summ 63 (Suppl 2):1–21
Devlin B, Scherer SW (2012) Genetic architecture in autism spectrum disorder. Curr Opin Genet Dev 22:229–237
Durand CM, Betancur C, Boeckers TM, Bockmann J, Chaste P, Fauchereau F, Nygren G, Rastam M, Gillberg IC, Anckarsäter H, Sponheim E, Goubran-Botros H, Delorme R, Chabane N, Mouren-Simeoni MC, de Mas P, Bieth E, Rogé B, Héron D, Burglen L, Gillberg C, Leboyer M, Bourgeron T (2007) Mutations in the gene encoding the synaptic scaffolding protein SHANK3 are associated with autism spectrum disorders. Nat Genet 39:25–27
Durand CM, Perroy J, Loll F, Perrais D, Fagni L, Bourgeron T, Montcouquiol M, Sans N (2012) SHANK3 mutations identified in autism lead to modification of dendritic spine morphology via an actin-dependent mechanism. Mol Psychiatry 17:71–84
Ebert DH, Gabel HW, Robinson ND, Kastan NR, Hu LS, Cohen S, Navarro AJ, Lyst MJ, Ekiert R, Bird AP, Greenberg ME (2013) Activity-dependent phosphorylation of MeCP2 threonine 308 regulates interaction with NCoR. Nature 499:341–345
Elsabbagh M, Divan G, Koh YJ, Kim YS, Kauchali S, Marcin C, Montiel-Nava C, Patel V, Paula CS, Wang C, Yasamy MT, Fombonne E (2012) Global prevalence of autism and other pervasive developmental disorders. Autism Res 5:160–179
Fernandez E, Rajan N, Bagni C (2013) The FMRP regulon: from targets to disease convergence. Front Neurosci 7:191
Folstein S, Rutter M (1977) Genetic influences and infantile autism. Nature 265:726–728
Fu W, O'Connor TD, Jun G, Kang HM, Abecasis G, Leal SM, Gabriel S, Rieder MJ, Altshuler D, Shendure J, Nickerson DA, Bamshad MJ, Project NES, Akey JM (2013) Analysis of 6,515 exomes reveals the recent origin of most human protein-coding variants. Nature 493:216–220
Gardener H, Spiegelman D, Buka SL (2011) Perinatal and neonatal risk factors for autism: a comprehensive meta-analysis. Pediatrics 128:344–355
Gaugler T, Klei L, Sanders SJ, Bodea CA, Goldberg AP, Lee AB, Mahajan M, Manaa D, Pawitan Y, Reichert J, Ripke S, Sandin S, Sklar P, Svantesson O, Reichenberg A, Hultman CM, Devlin B, Roeder K, Buxbaum JD (2014) Most genetic risk for autism resides with common variation. Nat Genet 46:881–885
Genome of the Netherlands C, Genome of the Netherlands C (2014) Whole-genome sequence variation, population structure and demographic history of the Dutch population. Nat Genet 46:818–825
Gillberg C (2010) The ESSENCE in child psychiatry: early symptomatic syndromes eliciting neurodevelopmental clinical examinations. Res Dev Disabil 31:1543–1551

Gillberg C, Wahlstrom J (1985) Chromosome abnormalities in infantile autism and other childhood psychoses: a population study of 66 cases. Dev Med Child Neurol 27:293–304

Girirajan S, Rosenfeld JA, Cooper GM, Antonacci F, Siswara P, Itsara A, Vives L, Walsh T, McCarthy SE, Baker C, Mefford HC, Kidd JM, Browning SR, Browning BL, Dickel DE, Levy DL, Ballif BC, Platky K, Farber DM, Gowans GC, Wetherbee JJ, Asamoah A, Weaver DD, Mark PR, Dickerson J, Garg BP, Ellingwood SA, Smith R, Banks VC, Smith W, McDonald MT, Hoo JJ, French BN, Hudson C, Johnson JP, Ozmore JR, Moeschler JB, Surti U, Escobar LF, El-Khechen D, Gorski JL, Kussmann J, Salbert B, Lacassie Y, Biser A, McDonald-McGinn DM, Zackai EH, Deardorff MA, Shaikh TH, Haan E, Friend KL, Fichera M, Romano C, Gécz J, DeLisi LE, Sebat J, King MC, Shaffer LG, Eichler EE (2010) A recurrent 16p12.1 microdeletion supports a two-hit model for severe developmental delay. Nat Genet 42:203–209

Girirajan S, Rosenfeld JA, Coe BP, Parikh S, Friedman N, Goldstein A, Filipink RA, McConnell JS, Angle B, Meschino WS, Nezarati MM, Asamoah A, Jackson KE, Gowans GC, Martin JA, Carmany EP, Stockton DW, Schnur RE, Penney LS, Martin DM, Raskin S, Leppig K, Thiese H, Smith R, Aberg E, Niyazov DM, Escobar LF, El-Khechen D, Johnson KD, Lebel RR, Siefkas K, Ball S, Shur N, McGuire M, Brasington CK, Spence JE, Martin LS, Clericuzio C, Ballif BC, Shaffer LG, Eichler EE (2012) Phenotypic heterogeneity of genomic disorders and rare copy-number variants. N Engl J Med 367:1321–1331

Gkogkas CG, Khoutorsky A, Ran I, Rampakakis E, Nevarko T, Weatherill DB, Vasuta C, Yee S, Truitt M, Dallaire P, Major F, Lasko P, Ruggero D, Nader K, Lacaille JC, Sonenberg N (2013) Autism-related deficits via dysregulated eIF4E-dependent translational control. Nature 493:371–377

Glessner JT, Wang K, Cai G, Korvatska O, Kim CE, Wood S, Zhang H, Estes A, Brune CW, Bradfield JP, Imielinski M, Frackelton EC, Reichert J, Crawford EL, Munson J, Sleiman PM, Chiavacci R, Annaiah K, Thomas K, Hou C, Glaberson W, Flory J, Otieno F, Garris M, Soorya L, Klei L, Piven J, Meyer KJ, Anagnostou E, Sakurai T, Game RM, Rudd DS, Zurawiecki D, McDougle CJ, Davis LK, Miller J, Posey DJ, Michaels S, Kolevzon A, Silverman JM, Bernier R, Levy SE, Schultz RT, Dawson G, Owley T, McMahon WM, Wassink TH, Sweeney JA, Nurnberger JI, Coon H, Sutcliffe JS, Minshew NJ, Grant SF, Bucan M, Cook EH, Buxbaum JD, Devlin B, Schellenberg GD, Hakonarson H (2009) Autism genome-wide copy number variation reveals ubiquitin and neuronal genes. Nature 459:569–573

Grabrucker AM, Schmeisser MJ, Schoen M, Boeckers TM (2011) Postsynaptic ProSAP/Shank scaffolds in the cross-hair of synaptopathies. Trends Cell Biol 21:594–603

Greer PL, Hanayama R, Bloodgood BL, Mardinly AR, Lipton DM, Flavell SW, Kim TK, Griffith EC, Waldon Z, Maehr R, Ploegh HL, Chowdhury S, Worley PF, Steen J, Greenberg ME (2010) The Angelman Syndrome protein Ube3A regulates synapse development by ubiquitinating arc. Cell 140:704–716

Gupta S, Ellis SE, Ashar FN, Moes A, Bader JS, Zhan J, West AB, Arking DE (2014) Transcriptome analysis reveals dysregulation of innate immune response genes and neuronal activity-dependent genes in autism. Nat Commun 5:5748

Hallmayer J, Cleveland S, Torres A, Phillips J, Cohen B, Torigoe T, Miller J, Fedele A, Collins J, Smith K, Lotspeich L, Croen LA, Ozonoff S, Lajonchere C, Grether JK, Risch N (2011) Genetic heritability and shared environmental factors among twin pairs with autism. Arch Gen Psychiatry 68:1095–1102

Hartman JL, Garvik B, Hartwell L (2001) Principles for the buffering of genetic variation. Science 291:1001–1004

He X, Sanders SJ, Liu L, De Rubeis S, Lim ET, Sutcliffe JS, Schellenberg GD, Gibbs RA, Daly MJ, Buxbaum JD, State MW, Devlin B, Roeder K (2013) Integrated model of de novo and inherited genetic variants yields greater power to identify risk genes. PLoS Genet 9:e1003671

Hormozdiari F, Penn O, Borenstein E, Eichler EE (2015) The discovery of integrated gene networks for autism and related disorders. Genome Res 25:142–154

Huguet G, Bourgeron T (2016) The genetic causes of autism spectrum disorders. In: Synaptic dysfunction in autism spectrum disorder and intellectual disability (in press)

Huguet G, Ey E, Bourgeron T (2013) The genetic landscapes of autism spectrum disorders. Annu Rev Genomics Hum Genet 14:191–213

Hung AY, Futai K, Sala C, Valtschanoff JG, Ryu J, Woodworth MA, Kidd FL, Sung CC, Miyakawa T, Bear MF, Weinberg RJ, Sheng M (2008) Smaller dendritic spines, weaker synaptic transmission, but enhanced spatial learning in mice lacking Shank1. J Neurosci 28:1697–1708

Iossifov I, Ronemus M, Levy D, Wang Z, Hakker I, Rosenbaum J, Yamrom B, Lee YH, Narzisi G, Leotta A, Kendall J, Grabowska E, Ma B, Marks S, Rodgers L, Stepansky A, Troge J, Andrews P, Bekritsky M, Pradhan K, Ghiban E, Kramer M, Parla J, Demeter R, Fulton LL, Fulton RS, Magrini VJ, Ye K, Darnell JC, Darnell RB, Mardis ER, Wilson RK, Schatz MC, McCombie WR, Wigler M (2012) De novo gene disruptions in children on the autistic spectrum. Neuron 74:285–299

Iossifov I, O'Roak BJ, Sanders SJ, Ronemus M, Krumm N, Levy D, Stessman HA, Witherspoon KT, Vives L, Patterson KE, Smith JD, Paeper B, Nickerson DA, Dea J, Dong S, Gonzalez LE, Mandell JD, Mane SM, Murtha MT, Sullivan CA, Walker MF, Waqar Z, Wei L, Willsey AJ, Yamrom B, Lee YH, Grabowska E, Dalkic E, Wang Z, Marks S, Andrews P, Leotta A, Kendall J, Hakker I, Rosenbaum J, Ma B, Rodgers L, Troge J, Narzisi G, Yoon S, Schatz MC, Ye K, McCombie WR, Shendure J, Eichler EE, State MW, Wigler M (2014) The contribution of de novo coding mutations to autism spectrum disorder. Nature 515:216–221

Jacquemont ML, Sanlaville D, Redon R, Raoul O, Cormier-Daire V, Lyonnet S, Amiel J, Le Merrer M, Heron D, de Blois MC, Prieur M, Vekemans M, Carter NP, Munnich A, Colleaux L, Philippe A (2006) Array-based comparative genomic hybridisation identifies high frequency of cryptic chromosomal rearrangements in patients with syndromic autism spectrum disorders. J Med Genet 43:843–849

Jamain S, Quach H, Betancur C, Rastam M, Colineaux C, Gillberg IC, Soderstrom H, Giros B, Leboyer M, Gillberg C, Bourgeron T (2003) Mutations of the X-linked genes encoding neuroligins NLGN3 and NLGN4 are associated with autism. Nat Genet 34:27–29

Jang SW, Liu X, Pradoldej S, Tosini G, Chang Q, Iuvone PM, Ye K (2010) N-acetylserotonin activates TrkB receptor in a circadian rhythm. Proc Natl Acad Sci USA 107:3876–3881

Jiang YH, Yuen RK, Jin X, Wang M, Chen N, Wu X, Ju J, Mei J, Shi Y, He M, Wang G, Liang J, Wang Z, Cao D, Carter MT, Chrysler C, Drmic IE, Howe JL, Lau L, Marshall CR, Merico D, Nalpathamkalam T, Thiruvahindrapuram B, Thompson A, Uddin M, Walker S, Luo J, Anagnostou E, Zwaigenbaum L, Ring RH, Wang J, Lajonchere C, Wang J, Shih A, Szatmari P, Yang H, Dawson G, Li Y, Scherer SW (2013) Detection of clinically relevant genetic variants in autism spectrum disorder by whole-genome sequencing. Am J Hum Genet 93:249–263

Kanner L (1943) Autistic disturbances of affective contact. Nerv Child 2:217–250

Kelleher RJ 3rd, Bear MF (2008) The autistic neuron: troubled translation? Cell 135:401–406

Klei L, Sanders SJ, Murtha MT, Hus V, Lowe JK, Willsey AJ, Moreno-De-Luca D, Yu TW, Fombonne E, Geschwind D, Grice DE, Ledbetter DH, Lord C, Mane SM, Martin CL, Martin DM, Morrow EM, Walsh CA, Melhem NM, Chaste P, Sutcliffe JS, State MW, Cook EH Jr, Roeder K, Devlin B (2012) Common genetic variants, acting additively, are a major source of risk for autism. Mol Autism 3:9

Kong A, Frigge ML, Masson G, Besenbacher S, Sulem P, Magnusson G, Gudjonsson SA, Sigurdsson A, Jonasdottir A, Wong WS, Sigurdsson G, Walters GB, Steinberg S, Helgason H, Thorleifsson G, Gudbjartsson DF, Helgason A, Magnusson OT, Thorsteinsdottir U, Stefansson K (2012) Rate of de novo mutations and the importance of father's age to disease risk. Nature 488:471–475

Krumm N, O'Roak BJ, Shendure J, Eichler EE (2014) A de novo convergence of autism genetics and molecular neuroscience. Trends Neurosci 37:95–105

Krumm N, Turner TN, Baker C, Vives L, Mohajeri K, Witherspoon K, Raja A, Coe BP, Stessman HA, He ZX, Leal SM, Bernier R, Eichler EE (2015) Excess of rare, inherited truncating mutations in autism. Nat Genet 47:582–588

Kumar RA, KaraMohamed S, Sudi J, Conrad DF, Brune C, Badner JA, Gilliam TC, Nowak NJ, Cook EH Jr, Dobyns WB, Christian SL (2008) Recurrent 16p11.2 microdeletions in autism. Hum Mol Genet 17:628–638

Lam KS, Aman MG, Arnold LE (2006) Neurochemical correlates of autistic disorder: a review of the literature. Res Dev Disabil 27:254–289

Le Couteur A, Bailey A, Goode S, Pickles A, Gottesman I, Robertson S, Rutter M (1996) A broader phenotype of autism: the clinical spectrum in twins. J Child Psychol Psychiatry 37:785–801

Leblond CS, Heinrich J, Delorme R, Proepper C, Betancur C, Huguet G, Konyukh M, Chaste P, Ey E, Rastam M, Anckarsäter H, Nygren G, Gillberg IC, Melke J, Toro R, Regnault B, Fauchereau F, Mercati O, Lemière N, Skuse D, Poot M, Holt R, Monaco AP, Järvelä I, Kantojärvi K, Vanhala R, Curran S, Collier DA, Bolton P, Chiocchetti A, Klauck SM, Poustka F, Freitag CM, Waltes R, Kopp M, Duketis E, Bacchelli E, Minopoli F, Ruta L, Battaglia A, Mazzone L, Maestrini E, Sequeira AF, Oliveira B, Vicente A, Oliveira G, Pinto D, Scherer SW, Zelenika D, Delepine M, Lathrop M, Bonneau D, Guinchat V, Devillard F, Assouline B, Mouren MC, Leboyer M, Gillberg C, Boeckers TM, Bourgeron T (2012) Genetic and functional analyses of SHANK2 mutations suggest a multiple hit model of autism spectrum disorders. PLoS Genet 8:e1002521

Leblond CS, Nava C, Polge A, Gauthier J, Huguet G, Lumbroso S, Giuliano F, Stordeur C, Depienne C, Mouzat K, Pinto D, Howe J, Lemière N, Durand CM, Guibert J, Ey E, Toro R, Peyre H, Mathieu A, Amsellem F, Rastam M, Gillberg IC, Rappold GA, Holt R, Monaco AP, Maestrini E, Galan P, Heron D, Jacquette A, Afenjar A, Rastetter A, Brice A, Devillard F, Assouline B, Laffargue F, Lespinasse J, Chiesa J, Rivier F, Bonneau D, Regnault B, Zelenika D, Delepine M, Lathrop M, Sanlaville D, Schluth-Bolard C, Edery P, Perrin L, Tabet AC, Schmeisser MJ, Boeckers TM, Coleman M, Sato D, Szatmari P, Scherer SW, Rouleau GA, Betancur C, Leboyer M, Gillberg C, Delorme R, Bourgeron T (2014) Meta-analysis of SHANK mutations in autism spectrum disorders: a gradient of severity in cognitive impairments. PLoS Genet 10:e1004580

Levy D, Ronemus M, Yamrom B, Lee YH, Leotta A, Kendall J, Marks S, Lakshmi B, Pai D, Ye K, Buja A, Krieger A, Yoon S, Troge J, Rodgers L, Iossifov I, Wigler M (2011) Rare de novo and transmitted copy-number variation in autistic spectrum disorders. Neuron 70:886–897

Lichtenstein P, Carlstrom E, Rastam M, Gillberg C, Anckarsater H (2010) The genetics of autism spectrum disorders and related neuropsychiatric disorders in childhood. Am J Psychiatry 167:1357–1363

Lim ET, Raychaudhuri S, Sanders SJ, Stevens C, Sabo A, MacArthur DG, Neale BM, Kirby A, Ruderfer DM, Fromer M, Lek M, Liu L, Flannick J, Ripke S, Nagaswamy U, Muzny D, Reid JG, Hawes A, Newsham I, Wu Y, Lewis L, Dinh H, Gross S, Wang LS, Lin CF, Valladares O, Gabriel SB, dePristo M, Altshuler DM, Purcell SM, NHLBI Exome Sequencing Project, State MW, Boerwinkle E, Buxbaum JD, Cook EH, Gibbs RA, Schellenberg GD, Sutcliffe JS, Devlin B, Roeder K, Daly MJ (2013) Rare complete knockouts in humans: population distribution and significant role in autism spectrum disorders. Neuron 77:235–242

Liu L, Sabo A, Neale BM, Nagaswamy U, Stevens C, Lim E, Bodea CA, Muzny D, Reid JG, Banks E, Coon H, Depristo M, Dinh H, Fennel T, Flannick J, Gabriel S, Garimella K, Gross S, Hawes A, Lewis L, Makarov V, Maguire J, Newsham I, Poplin R, Ripke S, Shakir K, Samocha KE, Wu Y, Boerwinkle E, Buxbaum JD, Cook EH Jr, Devlin B, Schellenberg GD, Sutcliffe JS, Daly MJ, Gibbs RA, Roeder K (2013) Analysis of rare, exonic variation amongst subjects with autism spectrum disorders and population controls. PLoS Genet 9:e1003443

Liu L, Lei J, Sanders SJ, Willsey AJ, Kou Y, Cicek AE, Klei L, Lu C, He X, Li M, Muhle RA, Ma'ayan A, Noonan JP, Sestan N, McFadden KA, State MW, Buxbaum JD, Devlin B, Roeder

K (2014) DAWN: a framework to identify autism genes and subnetworks using gene expression and genetics. Mol Autism 5:22

Lundstrom S, Chang Z, Kerekes N, Gumpert CH, Rastam M, Gillberg C, Lichtenstein P, Anckarsater H (2011) Autistic-like traits and their association with mental health problems in two nationwide twin cohorts of children and adults. Psychol Med 41:2423–2433

Ma XM, Blenis J (2009) Molecular mechanisms of mTOR-mediated translational control. Nat Rev Mol Cell Biol 10:307–318

Mabb AM, Ehlers MD (2010) Ubiquitination in postsynaptic function and plasticity. Annu Rev Cell Dev Biol 26:179–210

Malow B, Adkins KW, McGrew SG, Wang L, Goldman SE, Fawkes D, Burnette C (2012) Melatonin for sleep in children with autism: a controlled trial examining dose, tolerability, and outcomes. J Autism Dev Disord 42:1729–1737, author reply 1738

Maronde E, Saade A, Ackermann K, Goubran-Botros H, Pagan C, Bux R, Bourgeron T, Dehghani F, Stehle JH (2011) Dynamics in enzymatic protein complexes offer a novel principle for the regulation of melatonin synthesis in the human pineal gland. J Pineal Res 51:145–155

Melke J et al (2008) Abnormal melatonin synthesis in autism spectrum disorders. Mol Psychiatry 13:90–98

Michaelson JJ, Shi Y, Gujral M, Zheng H, Malhotra D, Jin X, Jian M, Liu G, Greer D, Bhandari A, Wu W, Corominas R, Peoples A, Koren A, Gore A, Kang S, Lin GN, Estabillo J, Gadomski T, Singh B, Zhang K, Akshoomoff N, Corsello C, McCarroll S, Iakoucheva LM, Li Y, Wang J, Sebat J (2012) Whole-genome sequencing in autism identifies hot spots for de novo germline mutation. Cell 151:1431–1442

Miles JH, Takahashi TN, Bagby S, Sahota PK, Vaslow DF, Wang CH, Hillman RE, Farmer JE (2005) Essential versus complex autism: definition of fundamental prognostic subtypes. Am J Med Genet A 135:171–180

Moreno-De-Luca A, Myers SM, Challman TD, Moreno-De-Luca D, Evans DW, Ledbetter DH (2013) Developmental brain dysfunction: revival and expansion of old concepts based on new genetic evidence. Lancet Neurol 12:406–414

Neale BM et al (2012) Patterns and rates of exonic de novo mutations in autism spectrum disorders. Nature 485:242–245

Nemirovsky SI, Cordoba M, Zaiat JJ, Completa SP, Vega PA, Gonzalez-Moron D, Medina NM, Fabbro M, Romero S, Brun B, Revale S, Ogara MF, Pecci A, Marti M, Vazquez M, Turjanski A, Kauffman MA (2015) Whole genome sequencing reveals a de novo SHANK3 mutation in familial autism spectrum disorder. PLoS One 10:e0116358

O'Roak BJ, Deriziotis P, Lee C, Vives L, Schwartz JJ, Girirajan S, Karakoc E, Mackenzie AP, Ng SB, Baker C, Rieder MJ, Nickerson DA, Bernier R, Fisher SE, Shendure J, Eichler EE (2011) Exome sequencing in sporadic autism spectrum disorders identifies severe de novo mutations. Nat Genet 43:585–589

O'Roak BJ, Vives L, Girirajan S, Karakoc E, Krumm N, Coe BP, Levy R, Ko A, Lee C, Smith JD, Turner EH, Stanaway IB, Vernot B, Malig M, Baker C, Reilly B, Akey JM, Borenstein E, Rieder MJ, Nickerson DA, Bernier R, Shendure J, Eichler EE (2012a) Sporadic autism exomes reveal a highly interconnected protein network of de novo mutations. Nature 485:246–250

O'Roak BJ, Vives L, Fu W, Egertson JD, Stanaway IB, Phelps IG, Carvill G, Kumar A, Lee C, Ankenman K, Munson J, Hiatt JB, Turner EH, Levy R, O'Day DR, Krumm N, Coe BP, Martin BK, Borenstein E, Nickerson DA, Mefford HC, Doherty D, Akey JM, Bernier R, Eichler EE, Shendure J (2012b) Multiplex targeted sequencing identifies recurrently mutated genes in autism spectrum disorders. Science 338:1619–1622

Obsil T, Ghirlando R, Klein DC, Ganguly S, Dyda F (2001) Crystal structure of the 14-3-3zeta: serotonin N-acetyltransferase complex. a role for scaffolding in enzyme regulation. Cell 105:257–267

Pagan C, Delorme R, Launay J, Bourgeron T (2012) Alterations of the serotonin-melatonin pathway in autism spectrum disorders: Biological evidence and clinical consequences. Oxford University Press, New York

Pagan C, Delorme R, Callebert J, Goubran-Botros H, Amsellem F, Drouot X, Boudebesse C, Le Dudal K, Ngo-Nguyen N, Laouamri H, Gillberg C, Leboyer M, Bourgeron T, Launay JM (2014) The serotonin-N-acetylserotonin-melatonin pathway as a biomarker for autism spectrum disorders. Transl Psychiatry 4:e479

Peca J, Feliciano C, Ting JT, Wang W, Wells MF, Venkatraman TN, Lascola CD, Fu Z, Feng G (2011) Shank3 mutant mice display autistic-like behaviours and striatal dysfunction. Nature 472:437–442

Pinto D, Pagnamenta AT, Klei L, Anney R, Merico D, Regan R, Conroy J, Magalhaes TR, Correia C, Abrahams BS, Almeida J, Bacchelli E, Bader GD, Bailey AJ, Baird G, Battaglia A, Berney T, Bolshakova N, Bölte S, Bolton PF, Bourgeron T, Brennan S, Brian J, Bryson SE, Carson AR, Casallo G, Casey J, Chung BH, Cochrane L, Corsello C, Crawford EL, Crossett A, Cytrynbaum C, Dawson G, de Jonge M, Delorme R, Drmic I, Duketis E, Duque F, Estes A, Farrar P, Fernandez BA, Folstein SE, Fombonne E, Freitag CM, Gilbert J, Gillberg C, Glessner JT, Goldberg J, Green A, Green J, Guter SJ, Hakonarson H, Heron EA, Hill M, Holt R, Howe JL, Hughes G, Hus V, Igliozzi R, Kim C, Klauck SM, Kolevzon A, Korvatska O, Kustanovich V, Lajonchere CM, Lamb JA, Laskawiec M, Leboyer M, Le Couteur A, Leventhal BL, Lionel AC, Liu XQ, Lord C, Lotspeich L, Lund SC, Maestrini E, Mahoney W, Mantoulan C, Marshall CR, McConachie H, McDougle CJ, McGrath J, McMahon WM, Merikangas A, Migita O, Minshew NJ, Mirza GK, Munson J, Nelson SF, Noakes C, Noor A, Nygren G, Oliveira G, Papanikolaou K, Parr JR, Parrini B, Paton T, Pickles A, Pilorge M, Piven J, Ponting CP, Posey DJ, Poustka A, Poustka F, Prasad A, Ragoussis J, Renshaw K, Rickaby J, Roberts W, Roeder K, Roge B, Rutter ML, Bierut LJ, Rice JP, Salt J, Sansom K, Sato D, Segurado R, Sequeira AF, Senman L, Shah N, Sheffield VC, Soorya L, Sousa I, Stein O, Sykes N, Stoppioni V, Strawbridge C, Tancredi R, Tansey K, Thiruvahindrapduram B, Thompson AP, Thomson S, Tryfon A, Tsiantis J, Van Engeland H, Vincent JB, Volkmar F, Wallace S, Wang K, Wang Z, Wassink TH, Webber C, Weksberg R, Wing K, Wittemeyer K, Wood S, Wu J, Yaspan BL, Zurawiecki D, Zwaigenbaum L, Buxbaum JD, Cantor RM, Cook EH, Coon H, Cuccaro ML, Devlin B, Ennis S, Gallagher L, Geschwind DH, Gill M, Haines JL, Hallmayer J, Miller J, Monaco AP, Nurnberger JI Jr, Paterson AD, Pericak-Vance MA, Schellenberg GD, Szatmari P, Vicente AM, Vieland VJ, Wijsman EM, Scherer SW, Sutcliffe JS, Betancur C (2010) Functional impact of global rare copy number variation in autism spectrum disorders. Nature 466:368–372

Pinto D, Darvishi K, Shi X, Rajan D, Rigler D, Fitzgerald T, Lionel AC, Thiruvahindrapuram B, Macdonald JR, Mills R, Prasad A, Noonan K, Gribble S, Prigmore E, Donahoe PK, Smith RS, Park JH, Hurles ME, Carter NP, Lee C, Scherer SW, Feuk L (2011) Comprehensive assessment of array-based platforms and calling algorithms for detection of copy number variants. Nat Biotechnol 29:512–520

Pinto D, Delaby E, Merico D, Barbosa M, Merikangas A, Klei L, Thiruvahindrapuram B, Xu X, Ziman R, Wang Z, Vorstman JA, Thompson A, Regan R, Pilorge M, Pellecchia G, Pagnamenta AT, Oliveira B, Marshall CR, Magalhaes TR, Lowe JK, Howe JL, Griswold AJ, Gilbert J, Duketis E, Dombroski BA, De Jonge MV, Cuccaro M, Crawford EL, Correia CT, Conroy J, Conceição IC, Chiocchetti AG, Casey JP, Cai G, Cabrol C, Bolshakova N, Bacchelli E, Anney R, Gallinger S, Cotterchio M, Casey G, Zwaigenbaum L, Wittemeyer K, Wing K, Wallace S, van Engeland H, Tryfon A, Thomson S, Soorya L, Rogé B, Roberts W, Poustka F, Mouga S, Minshew N, McInnes LA, McGrew SG, Lord C, Leboyer M, Le Couteur AS, Kolevzon A, Jiménez González P, Jacob S, Holt R, Guter S, Green J, Green A, Gillberg C, Fernandez BA, Duque F, Delorme R, Dawson G, Chaste P, Café C, Brennan S, Bourgeron T, Bolton PF, Bölte S, Bernier R, Baird G, Bailey AJ, Anagnostou E, Almeida J, Wijsman EM, Vieland VJ, Vicente AM, Schellenberg GD, Pericak-Vance M, Paterson AD, Parr JR, Oliveira G, Nurnberger JI, Monaco AP, Maestrini E, Klauck SM, Hakonarson H, Haines JL,

Geschwind DH, Freitag CM, Folstein SE, Ennis S, Coon H, Battaglia A, Szatmari P, Sutcliffe JS, Hallmayer J, Gill M, Cook EH, Buxbaum JD, Devlin B, Gallagher L, Betancur C, Scherer SW (2014) Convergence of genes and cellular pathways dysregulated in autism spectrum disorders. Am J Hum Genet 94:677–694

Risch N, Hoffmann TJ, Anderson M, Croen LA, Grether JK, Windham GC (2014) Familial recurrence of autism spectrum disorder: evaluating genetic and environmental contributions. Am J Psychiatry 171:1206–1213

Ritvo ER, Freeman BJ, Mason-Brothers A, Mo A, Ritvo AM (1985) Concordance for the syndrome of autism in 40 pairs of afflicted twins. Am J Psychiatry 142:74–77

Ronald A, Hoekstra RA (2011) Autism spectrum disorders and autistic traits: a decade of new twin studies. Am J Med Genet B Neuropsychiatr Genet 156B:255–274

Ronald A, Happe F, Plomin R (2005) The genetic relationship between individual differences in social and nonsocial behaviours characteristic of autism. Dev Sci 8:444–458

Ronald A, Happe F, Price TS, Baron-Cohen S, Plomin R (2006) Phenotypic and genetic overlap between autistic traits at the extremes of the general population. J Am Acad Child Adolesc Psychiatry 45:1206–1214

Ronald A, Larsson H, Anckarsater H, Lichtenstein P (2010) A twin study of autism symptoms in Sweden. Mol Psychiatry 16:1039–1047

Ronemus M, Iossifov I, Levy D, Wigler M (2014) The role of de novo mutations in the genetics of autism spectrum disorders. Nat Rev Genet 15:133–141

Rosenberg RE, Law JK, Yenokyan G, McGready J, Kaufmann WE, Law PA (2009) Characteristics and concordance of autism spectrum disorders among 277 twin pairs. Arch Pediatr Adolesc Med 163:907–914

Rossignol DA, Frye RE (2012) A review of research trends in physiological abnormalities in autism spectrum disorders: immune dysregulation, inflammation, oxidative stress, mitochondrial dysfunction and environmental toxicant exposures. Mol Psychiatry 17:389–401

Rutherford SL (2000) From genotype to phenotype: buffering mechanisms and the storage of genetic information. Bioessays 22:1095–1105

Samocha KE, Robinson EB, Sanders SJ, Stevens C, Sabo A, McGrath LM, Kosmicki JA, Rehnström K, Mallick S, Kirby A, Wall DP, MacArthur DG, Gabriel SB, DePristo M, Purcell SM, Palotie A, Boerwinkle E, Buxbaum JD, Cook EH Jr, Gibbs RA, Schellenberg GD, Sutcliffe JS, Devlin B, Roeder K, Neale BM, Daly MJ (2014) A framework for the interpretation of de novo mutation in human disease. Nat Genet 46:944–950

Sanders SJ, Ercan-Sencicek AG, Hus V, Luo R, Murtha MT, Moreno-De-Luca D, Chu SH, Moreau MP, Gupta AR, Thomson SA, Mason CE, Bilguvar K, Celestino-Soper PB, Choi M, Crawford EL, Davis L, Wright NR, Dhodapkar RM, DiCola M, DiLullo NM, Fernandez TV, Fielding-Singh V, Fishman DO, Frahm S, Garagaloyan R, Goh GS, Kammela S, Klei L, Lowe JK, Lund SC, McGrew AD, Meyer KA, Moffat WJ, Murdoch JD, O'Roak BJ, Ober GT, Pottenger RS, Raubeson MJ, Song Y, Wang Q, Yaspan BL, Yu TW, Yurkiewicz IR, Beaudet AL, Cantor RM, Curland M, Grice DE, Günel M, Lifton RP, Mane SM, Martin DM, Shaw CA, Sheldon M, Tischfield JA, Walsh CA, Morrow EM, Ledbetter DH, Fombonne E, Lord C, Martin CL, Brooks AI, Sutcliffe JS, Cook EH Jr, Geschwind D, Roedr K, Devlin B, State MW (2011) Multiple recurrent de novo CNVs, including duplications of the 7q11.23 Williams syndrome region, are strongly associated with autism. Neuron 70:863–885

Sanders SJ, Murtha MT, Gupta AR, Murdoch JD, Raubeson MJ, Willsey AJ, Ercan-Sencicek AG, DiLullo NM, Parikshak NN, Stein JL, Walker MF, Ober GT, Teran NA, Song Y, El-Fishawy P, Murtha RC, Choi M, Overton JD, Bjornson RD, Carriero NJ, Meyer KA, Bilguvar K, Mane SM, Sestan N, Lifton RP, Günel M, Roeder K, Geschwind DH, Devlin B, State MW (2012) De novo mutations revealed by whole-exome sequencing are strongly associated with autism. Nature 485:237–241

Sandin S, Lichtenstein P, Kuja-Halkola R, Larsson H, Hultman CM, Reichenberg A (2014) The familial risk of autism. JAMA 311:1770–1777

Sando R 3rd, Gounko N, Pieraut S, Liao L, Yates J 3rd, Maximov A (2012) HDAC4 governs a transcriptional program essential for synaptic plasticity and memory. Cell 151:821–834

Santini E, Huynh TN, MacAskill AF, Carter AG, Pierre P, Ruggero D, Kaphzan H, Klann E (2013) Exaggerated translation causes synaptic and behavioural aberrations associated with autism. Nature 493:411–415

Schmeisser MJ, Ey E, Wegener S, Bockmann J, Stempel AV, Kuebler A, Janssen AL, Udvardi PT, Shiban E, Spilker C, Balschun D, Skryabin BV, Dieck S, Smalla KH, Montag D, Leblond CS, Faure P, Torquet N, Le Sourd AM, Toro R, Grabrucker AM, Shoichet SA, Schmitz D, Kreutz MR, Bourgeron T, Gundelfinger ED, Boeckers TM (2012) Autistic-like behaviours and hyperactivity in mice lacking ProSAP1/Shank2. Nature 486:256–260

Sebat J, Lakshmi B, Malhotra D, Troge J, Lese-Martin C, Walsh T, Yamrom B, Yoon S, Krasnitz A, Kendall J, Leotta A, Pai D, Zhang R, Lee YH, Hicks J, Spence SJ, Lee AT, Puura K, Lehtimäki T, Ledbetter D, Gregersen PK, Bregman J, Sutcliffe JS, Jobanputra V, Chung W, Warburton D, King MC, Skuse D, Geschwind DH, Gilliam TC, Ye K, Wigler M (2007) Strong association of de novo copy number mutations with autism. Science 316:445–449

Sheng M, Kim E (2011) The postsynaptic organization of synapses. Cold Spring Harb Perspect Biol 3

Shi L, Zhang X, Golhar R, Otieno FG, He M, Hou C, Kim C, Keating B, Lyon GJ, Wang K, Hakonarson H (2013) Whole-genome sequencing in an autism multiplex family. Mol Autism 4:8

Silverman JL, Turner SM, Barkan CL, Tolu SS, Saxena R, Hung AY, Sheng M, Crawley JN (2011) Sociability and motor functions in Shank1 mutant mice. Brain Res 1380:120–137

Skuse DH, Mandy W, Steer C, Miller LL, Goodman R, Lawrence K, Emond A, Golding J (2009) Social communication competence and functional adaptation in a general population of children: preliminary evidence for sex-by-verbal IQ differential risk. J Am Acad Child Adolesc Psychiatry 48:128–137

Sompol P, Liu X, Baba K, Paul KN, Tosini G, Iuvone PM, Ye K (2011) N-acetylserotonin promotes hippocampal neuroprogenitor cell proliferation in sleep-deprived mice. Proc Natl Acad Sci USA 108:8844–8849

Steffenburg S, Gillberg C, Hellgren L, Andersson L, Gillberg IC, Jakobsson G, Bohman M (1989) A twin study of autism in Denmark, Finland, Iceland, Norway and Sweden. J Child Psychol Psychiat All Discip 30:405–416

Szafranski P, Schaaf CP, Person RE, Gibson IB, Xia Z, Mahadevan S, Wiszniewska J, Bacino CA, Lalani S, Potocki L, Kang SH, Patel A, Cheung SW, Probst FJ, Graham BH, Shinawi M, Beaudet AL, Stankiewicz P (2010) Structures and molecular mechanisms for common 15q13.3 microduplications involving CHRNA7: benign or pathological? Hum Mutat 31:840–850

Szatmari P, Paterson AD, Zwaigenbaum L, Roberts W, Brian J, Liu XQ, Vincent JB, Skaug JL, Thompson AP, Senman L, Feuk L, Qian C, Bryson SE, Jones MB, Marshall CR, Scherer SW, Vieland VJ, Bartlett C, Mangin LV, Goedken R, Segre A, Pericak-Vance MA, Cuccaro ML, Gilbert JR, Wright HH, Abramson RK, Betancur C, Bourgeron T, Gillberg C, Leboyer M, Buxbaum JD, Davis KL, Hollander E, Silverman JM, Hallmayer J, Lotspeich L, Sutcliffe JS, Haines JL, Folstein SE, Piven J, Wassink TH, Sheffield V, Geschwind DH, Bucan M, Brown WT, Cantor RM, Constantino JN, Gilliam TC, Herbert M, Lajonchere C, Ledbetter DH, Lese-Martin C, Miller J, Nelson S, Samango-Sprouse CA, Spence S, State M, Tanzi RE, Coon H, Dawson G, Devlin B, Estes A, Flodman P, Klei L, McMahon WM, Minshew N, Munson J, Korvatska E, Rodier PM, Schellenberg GD, Smith M, Spence MA, Stodgell C, Tepper PG, Wijsman EM, Yu CE, Rogé B, Mantoulan C, Wittemeyer K, Poustka A, Felder B, Klauck SM, Schuster C, Poustka F, Bölte S, Feineis-Matthews S, Herbrecht E, Schmötzer G, Tsiantis J, Papanikolaou K, Maestrini E, Bacchelli E, Blasi F, Carone S, Toma C, Van Engeland H, de Jonge M, Kemner C, Koop F, Langemeijer M, Hijmans C, Staal WG, Baird G, Bolton PF, Rutter ML, Weisblatt E, Green J, Aldred C, Wilkinson JA, Pickles A, Le Couteur A, Berney T, McConachie H, Bailey AJ, Francis K, Honeyman G, Hutchinson A, Parr JR, Wallace S,

Monaco AP, Barnby G, Kobayashi K, Lamb JA, Sousa I, Sykes N, Cook EH, Guter SJ, Leventhal BL, Salt J, Lord C, Corsello C, Hus V, Weeks DE, Volkmar F, Tauber M, Fombonne E, Shih A, Meyer KJ (2007) Mapping autism risk loci using genetic linkage and chromosomal rearrangements. Nat Genet 39:319–328

Taniai H, Nishiyama T, Miyachi T, Imaeda M, Sumi S (2008) Genetic influences on the broad spectrum of autism: study of proband-ascertained twins. Am J Med Genet B Neuropsychiatr Genet 147B:844–849

Tordjman S, Anderson GM, Pichard N, Charbuy H, Touitou Y (2005) Nocturnal excretion of 6-sulphatoxymelatonin in children and adolescents with autistic disorder. Biol Psychiatry 57:134–138

Toro R, Konyukh M, Delorme R, Leblond C, Chaste P, Fauchereau F, Coleman M, Leboyer M, Gillberg C, Bourgeron T (2010) Key role for gene dosage and synaptic homeostasis in autism spectrum disorders. Trends Genet 26:363–372

Toro R, Poline JB, Huguet G, Loth E, Frouin V, Banaschewski T, Barker GJ, Bokde A, Büchel C, CarvalhoFM CP, Fauth-Bühler M, Flor H, Gallinat J, Garavan H, Gowland P, Heinz A, Ittermann B, Lawrence C, Lemaître H, Mann K, Nees F, Paus T, Pausova Z, Rietsche M, Robbins T, Smolka MN, Ströhle A, Schumann G, Bourgeron T (2014) Genomic architecture of human neuroanatomical diversity. Mol Psychiatry 20:1011–1016

Uddin M, Tammimies K, Pellecchia G, Alipanahi B, Hu P, Wang Z, Pinto D, Lau L, Nalpathamkalam T, Marshall CR, Blencowe BJ, Frey BJ, Merico D, Yuen RK, Scherer SW (2014) Brain-expressed exons under purifying selection are enriched for de novo mutations in autism spectrum disorder. Nat Genet 46:742–747

Voineagu I, Wang X, Johnston P, Lowe JK, Tian Y, Horvath S, Mill J, Cantor RM, Blencowe BJ, Geschwind DH (2011) Transcriptomic analysis of autistic brain reveals convergent molecular pathology. Nature 474:380–384

Vorstman JA, Staal WG, van Daalen E, van Engeland H, Hochstenbach PF, Franke L (2006) Identification of novel autism candidate regions through analysis of reported cytogenetic abnormalities associated with autism. Mol Psychiatry 11(1):18–28

Wang X, McCoy PA, Rodriguiz RM, Pan Y, Je HS, Roberts AC, Kim CJ, Berrios J, Colvin JS, Bousquet-Moore D, Lorenzo I, Wu G, Weinberg RJ, Ehlers MD, Philpot BD, Beaudet AL, Wetsel WC, Jiang YH (2011) Synaptic dysfunction and abnormal behaviors in mice lacking major isoforms of Shank3. Hum Mol Genet 20:3093–3108

Weiss LA, Shen Y, Korn JM, Arking DE, Miller DT, Fossdal R, Saemundsen E, Stefansson H, Ferreira MA, Green T, Platt OS, Ruderfer DM, Walsh CA, Altshuler D, Chakravarti A, Tanzi RE, Stefansson K, Santangelo SL, Gusella JF, Sklar P, Wu BL, Daly MJ, Autism Consortium (2008) Association between microdeletion and microduplication at 16p11.2 and autism. N Engl J Med 358:667–675

Willsey AJ, Sanders SJ, Li M, Dong S, Tebbenkamp AT, Muhle RA, Reilly SK, Lin L, Fertuzinhos S, Miller JA, Murtha MT, Bichsel C, Niu W, Cotney J, Ercan-Sencicek AG, Gockley J, Gupta AR, Han W, He X, Hoffman EJ, Klei L, Lei J, Liu W, Liu L, Lu C, Xu X, Zhu Y, Mane SM, Lein ES, Wei L, Noonan JP, Roeder K, Devlin B, Sestan N, State MW (2013) Coexpression networks implicate human midfetal deep cortical projection neurons in the pathogenesis of autism. Cell 155:997–1007

Wohr M, Roullet FI, Hung AY, Sheng M, Crawley JN (2011) Communication impairments in mice lacking Shank1: reduced levels of ultrasonic vocalizations and scent marking behavior. PLoS One 6:e20631

Won H, Lee H-R, Gee HY, Mah W, Kim J-I, Lee J, Ha S, Chung C, Jung ES, Cho YS, Park S-G, Lee J-S, Lee K, Kim D, Bae YC, Kaang B-K, Lee MG, Kim E (2012) Autistic-like social behaviour in Shank2-mutant mice improved by restoring NMDA receptor function. Nature 486:261–265

Wright B, Sims D, Smart S, Alwazeer A, Alderson-Day B, Allgar V, Whitton C, Tomlinson H, Bennett S, Jardine J, McCaffrey N, Leyland C, Jakeman C, Miles J (2011) Melatonin versus placebo in children with autism spectrum conditions and severe sleep problems not amenable to behaviour management strategies: a randomised controlled crossover trial. J Autism Dev Disord 41:175–184

Xue Y, Chen Y, Ayub Q, Huang N, Ball EV, Mort M, Phillips AD, Shaw K, Stenson PD, Cooper DN, Tyler-Smith C, Genomes Project C (2012) Deleterious- and disease-allele prevalence in healthy individuals: insights from current predictions, mutation databases, and population-scale resequencing. Am J Hum Genet 91:1022–1032

Yang J, Benyamin B, McEvoy BP, Gordon S, Henders AK, Nyholt DR, Madden PA, Heath AC, Martin NG, Montgomery GW, Goddard ME, Visscher PM (2010) Common SNPs explain a large proportion of the heritability for human height. Nat Genet 42:565–569

Yang J, Lee SH, Goddard ME, Visscher PM (2011) GCTA: a tool for genome-wide complex trait analysis. Am J Hum Genet 88:76–82

Yang M, Bozdagi O, Scattoni ML, Woehr M, Roullet FI, Katz AM, Abrams DN, Kalikhman D, Simon H, Woldeyohannes L, Zhang JY, Harris MJ, Saxena R, Silverman JL, Buxbaum JD, Crawley JN (2012) Reduced excitatory neurotransmission and mild autism-relevant phenotypes in adolescent Shank3 null mutant mice. J Neurosci 32:6525–6541

Yu TW, Chahrour MH, Coulter ME, Jiralerspong S, Okamura-Ikeda K, Ataman B, Schmitz-Abe K, Harmin DA, Adli M, Malik AN, D'Gama AM, Lim ET, Sanders SJ, Mochida GH, Partlow JN, Sunu CM, Felie JM, Rodriguez J, Nasir RH, Ware J, Joseph RM, Hill RS, Kwan BY, Al-Saffar M, Mukaddes NM, Hashmi A, Balkhy S, Gascon GG, Hisama FM, LeClair E, Poduri A, Oner O, Al-Saad S, Al-Awadi SA, Bastaki L, Ben-Omran T, Teebi AS, Al-Gazali L, Eapen V, Stevens CR, Rappaport L, Gabriel SB, Markianos K, State MW, Greenberg ME, Taniguchi H, Braverman NE, Morrow EM, Walsh CA (2013) Using whole-exome sequencing to identify inherited causes of autism. Neuron 77:259–273

Yuen RK, Thiruvahindrapuram B, Merico D, Walker S, Tammimies K, Hoang N, Chrysler C, Nalpathamkalam T, Pellecchia G, Liu Y, Gazzellone MJ, D'Abate L, Deneault E, Howe JL, Liu RS, Thompson A, Zarrei M, Uddin M, Marshall CR, Ring RH, Zwaigenbaum L, Ray PN, Weksberg R, Carter MT, Fernandez BA, Roberts W, Szatmari P, Scherer SW (2015) Whole-genome sequencing of quartet families with autism spectrum disorder. Nat Med 21:185–191

Index

© The Author(s) 2016
P. Sassone-Corsi, Y. Christen (eds.), *A Time for Metabolism and Hormones*,
Research and Perspectives in Endocrine Interactions,
DOI 10.1007/978-3-319-27069-2